Linear Estimation and Stochastic Control

D1545059

CHAPMAN AND HALL
MATHEMATICS SERIES

Edited by Professor R. Brown
Head of the Department of Pure Mathematics,
University College of North Wales, Bangor
and Dr. M. A. H. Dempster,
University Lecturer in Industrial Mathematics
and Fellow of Balliol College, Oxford

A Preliminary Course in Analysis
R. M. F. Moss and G. T. Roberts

Elementary Differential Equations
R. L. E. Schwarzenberger

A First Course on Complex Functions
G. J. O. Jameson

Rings, Modules and Linear Algebra
B. Hartley and T. O. Hawkes

Regular Algebra and Finite Machines
J. H. Conway

Complex Numbers
W. H. Cockcroft

Galois Theory
Ian Stewart

Topology and Normed Spaces
G. J. O. Jameson

Introduction to Optimization Methods
P. R. Adby and M. A. H. Dempster

Graphs, Surfaces and Homology
P. J. Giblin

Davis, M. H. A.
Linear estimation and stochastic
control / M. H. A. Davis. London :
Chapman and Hall ; New York : Wiley :
distributed in the U.S.A. by Halsted
Press, 1977.
xii, 224 p. : ill. ; 23 cm. (Chapman
and Hall mathematics series)

Linear Estimation and Stochastic Control

M. H. A. DAVIS

Department of Computing and Control,
Imperial College,
University of London

LONDON
CHAPMAN AND HALL

A Halsted Press Book
John Wiley & Sons, New York

329121

First published 1977
by Chapman and Hall Ltd
11 New Fetter Lane, London EC4P 4EE

© 1977 M. H. A. Davis

Typeset by The Alden Press (London and Northampton) Ltd
Printed in Great Britain at the University Printing House, Cambridge

ISBN 0 412 15470 6 (cased edition)

ISBN 0 412 15130 8 (paperback edition)

Distributed in the U.S.A. by Halsted Press,
a Division of John Wiley & Sons, Inc., New York

Library of Congress Cataloging in Publication Data

Davies, M   H   A
  Linear estimation and stochastic control.

  (Chapman and Hall mathematics series)
  "A Halsted Press book."
  Bibliography: p
  Includes index.
  1. Control theory. 2. Estimation theory.
3. Stochastic processes. I. Title.
QA402.3.D38   519.2   77-23389
ISBN 0-470-99215-8

Contents

Preface

This book deals with problems of estimation and control in linear stochastic dynamical systems, i.e. systems represented by linear vector differential equations of the form

$$\dot{x}_t = Ax_t + Bu(t) + Cw_t$$

where x_t is the state vector, $u(t)$ a control input and w_t a random disturbance function. The types of disturbances considered are 'white' (i.e. wide band) noise and processes which can be generated from it, since only for such processes does x_t (or some augmented vector) retain the 'state' property that the evolution of the system beyond time t depends solely on the value x_t and on future inputs and disturbances.

Mathematically, linear estimation is projection in Hilbert space and white noise is represented by orthogonal increments processes. The first three chapters study these ideas (as well as providing a general introduction to stochastic processes) and explore the connection between Hilbert subspaces and Wiener integrals. Chapters 4 and 5 cover the main results in linear stochastic system theory, namely the Kalman filter, quadratic cost control problems and the certainty–equivalence principle. For the Kalman filter I follow the 'innovations approach' introduced by Kailath in his seminal paper [32]. This exposes the structure of the problem far better than rival methods and is indeed the basis for all further developments in filtering theory. On the control side the main though less direct influence has been Wonham [20, 30, 39]. Again, the arguments parallel as closely as possible those used in non-linear problems.

These topics have been described by Balakrishnan [11] as the bread-and-butter part of stochastic control theory. The purpose of Chapter 6 is to give just a taste of the cakes and pastries.

It should be mentioned that the alternative theory of linear filtering, formulated in the context of stationary processes and leading via spectral analysis to the Wiener filter, is not included, and for this reason, although 'filtering' has almost lost its frequency-domain connotation, I have tended to prefer the term 'estimation', even at the risk of offending some non-Bayesians.

My objective has been to write this book at the precise mathematical level required for a rigorous formulation of the main results. The level is fixed by the central idea of estimation as projection in Hilbert space; this means that some measure theory is inescapable. However, this only intervenes in connection with technical points, the significance of which can be appreciated without following through their proofs. Thus I have two categories of reader in mind. The basic prerequisites are a familiarity with elementary probability theory at the level of, say, Larson [5], some linear system theory [22, 24] and the elements of real analysis; readers with this background will be able to follow everything except the Appendix and some proofs, marked with an asterisk, the omission of which will not result in any loss of continuity. These sections will, however, be accessible to those with a knowledge of measure theory equivalent to about the first five chapters of Kingman and Taylor [4], to which reference is made for standard results in this subject. Readers in either category are advised to deal cursorily with the more technical parts of Chapter 2 until it becomes apparent in later sections why such results are required. Some of the Problems are stated in the form of an outline argument, the 'problem' being of course to fill in the details.

I am indebted to Martin Clark and Richard Vinter for helpful discussions and to several other colleagues whose comments have eliminated some errors and obscurities. The typing was done mainly by Joan Jelinek and Christine Ware, though Magda Bicknell and Linden Rice also had a hand in it; their friendly collaboration made the task of writing this book much less onerous than it otherwise could have been.

London, 1977   M. H. A. Davis

Abbreviations and Notation

Abbreviations used in the text

a.s.	almost surely
BM	Brownian motion
d.f.	distribution function
i.i.d.	independent, identically distributed
o.i.	orthogonal increments
o.n.	orthonormal
q.m.	quadratic mean
r.v.	random variable

A function f with domain X and range Y is denoted $f: X \to Y$ or $f(\cdot)$ or just f if X and Y are not in doubt. $f(x) \in Y$ is the value of f at $x \in X$. If A is any subset of X, the *indicator function* of A is the function $I_A : X \to \{0, 1\}$ defined by $I_A = 1$ if $x \in A$, $I_A = 0$ if $x \notin A$. $\{x \in X : P(x)\}$ is the subset of X consisting of those elements having property P. For a probability space Ω (see Section 2.1) it is customary to abbreviate this notation so that for example $\{\omega \in \Omega : Z(\omega) < a\}$ becomes simply $(Z < a)$.

The real line is denoted by R and the intervals $\{x \in R : a \leqslant x < b\}$ by $[a, b)$ (and analogously for $[a, b]$, etc.). For $s, t \in R$, $s \wedge t$ is the lesser of t and s, i.e.

$$t \wedge s = \tfrac{1}{2}(t + s - |t - s|).$$

Euclidean n-space is $R^n = \{(x_1, x_2 \ldots x_n): x_i \in R, i = 1,$
$2 \ldots n\}$ with inner product $x \cdot y = \sum_i x_i y_i$ for $x = (x_1 \ldots x_n)$
and $y = (y_1 \ldots y_n)$. A' is the transpose of a matrix A; thus
$x \cdot y = x'y$ for $x, y \in R^n$.

If X and Y are r.v.'s, EX is the *expectation* of X, defined
by equation (2.6). Then the *covariance* of X and Y and the
variance of X are given respectively by

$$\mathrm{cov}(X, Y) = E[(X - EX)(Y - EY)] = EXY - EX \cdot EY$$

$$\mathrm{var}(X) = \mathrm{cov}(X, X).$$

For a vector r.v. $X' = (X_1, X_2 \ldots X_n)$, the *covariance
matrix* $\mathrm{cov}(X)$ is the $n \times n$ matrix whose i, jth element is
$\mathrm{cov}(X_i, X_j)$, i.e.

$$\mathrm{cov}(X) = EXX' - (EX)(EX)'.$$

$N(\mu, \sigma^2)$ denotes the normal distribution with expectation μ
and variance σ^2.

A stochastic process $\{X_t, t \in T\}$ (see Definition 2.1.2) will
be written $\{X_t\}$ if the identity of the time set T is clear. X_t is
the r.v. corresponding to a fixed time $t \in T$ and $X_t(\omega)$ the
value of X_t at a particular elementary event $\omega \in \Omega$.

Generally a process $\{X_t\}$ denoted by an upper case letter is
scalar-valued whereas a lower-case process $\{x_t\}$ is vector-
valued. An exception to this is that the ith component of a
vector process $\{x_t\}$ is written $\{x_t^i\}$.

CHAPTER ONE

Finite-dimensional linear estimation

1.1. The geometrical structure of linear estimation

The objective of this chapter is to introduce the 'geometric' structure of linear estimation problems by considering the problem of estimating a random variable (r.v.) X given the value of a related n-vector r.v. Y. It will be seen that linear estimation amounts to projection in R^{n+1}. The results will then be generalized in Chapter 2 to estimation of given *processes*, i.e. more than just a finite number of r.v.'s. But since this involves a lot more mathematical machinery, it seems advantageous to treat the finite-dimensional case in a self-contained way first.

Suppose $Y_1, Y_2 \ldots Y_n$ are independent r.v.'s with

$$EY_i = 0, \qquad \mathrm{var}(Y_i) = EY_i^2 < \infty$$

and let X be another zero-mean, finite-variance r.v. A *linear estimator* of X given $Y_1 \ldots Y_n$ is any linear combination

$$\hat{X} = \sum_i \alpha_i Y_i \qquad (1.1)$$

and the mean-square error is $E(X - \hat{X})^2$. How should the α_i be chosen so as to minimize this? Since $EY_iY_j = 0$ for $i \neq j$, we have

$$E\left(X - \sum_i \alpha_i Y_i\right)^2 = EX^2 + \sum_i \alpha_i^2 EY_i^2 - 2\sum_i \alpha_i E(Y_i X).$$

1

Thus

$$\frac{\partial}{\partial \alpha_i} E(X - \hat{X})^2 = 2\alpha_i EY_i^2 - 2E(Y_iX).$$

It follows that $E(X - \hat{X})^2$ is minimized when

$$\alpha_i = \frac{E(Y_iX)}{E(Y_i^2)}. \tag{1.2}$$

This result can be reinterpreted in geometrical terms in the following way. Let $\mathcal{H}$ be the set of all linear combinations of $Y_1 \ldots Y_n$ and X, i.e.

$$\mathcal{H} = \left\{ \sum_{i=1}^{n} \beta_i Y_i + \beta_{n+1} X \,\middle|\, \beta \in R^{n+1} \right\}$$

For $U, V \in \mathcal{H}$ define the *inner product*

$$(U, V) = EUV$$

and the *norm*

$$\|U\| = \sqrt{(U, U)}$$

U and V are said to be orthogonal, $U \perp V$, if $(U, V) = 0$.
For $U_1 \ldots U_k \in \mathcal{H}$, the *subspace spanned* by $U_1 \ldots U_k$ is

$$\mathcal{L}(U_1 \ldots U_k) = \left\{ \sum_{i=1}^{k} \beta_i U_i \,\middle|\, \beta \in R^k \right\}$$

We say $V \perp \mathcal{L}(U_1 \ldots U_k)$ if $V \perp U$ for all $U \in \mathcal{L}(U_1 \ldots U_k)$. Now let $\hat{X}$ be the linear least squares estimator given by (1.1) and (1.2).

1.1.1. Proposition. $\hat{X}$ *is characterized by*

(a) $\hat{X} \in \mathcal{L}(Y_1 \ldots Y_n)$

(b) $X - \hat{X} \perp \mathcal{L}(Y_1 \ldots Y_n)$

Proof. Suppose $Z \in \mathcal{L}(Y_1 \ldots Y_n)$. Then

$$Z = \sum_{i=1}^{n} \beta_i Y_i$$

for some $\beta \in R^n$. So $(X - Z) \perp \mathcal{L}(Y_1 \ldots Y_n)$ if and only if

$(X - Z) \perp Y_i$ for all i, which is equivalent to saying

$$(X, Y_i) = (Z, Y_i) = \beta_i(Y_i, Y_i)$$

i.e.

$$\beta_i = \alpha_i.$$

But then $Z = \hat{X}$. This completes the proof.

$\hat{X}$ satisfying (a) and (b) above is called the *orthogonal projection* of X onto $\mathcal{L}(Y_1 \ldots Y_n)$.

Notice that all the above calculations only involve the means and covariances of the r.v.'s. In particular it would have been sufficient to assume that the Y_i's were merely uncorrelated rather than independent. Notice also that 'uncorrelated' and 'orthogonal' are synonymous in the present context. We now show that the characterization of $\hat{X}$ given in Proposition 1.1.1 does not in fact depend on the mutual orthogonality of the Y_i's, because if they are not orthogonal we can construct an equivalent orthogonal set by the 'Gram–Schmidt' procedure described below.

1.1.2. Proposition. *Let* $Y_1 \ldots Y_n$ *be zero-mean, finite variance r.v.'s. Then there exists an integer* $m \leqslant n$ *and random variables* $Z_1 \ldots Z_m$ *such that*

(a) $\|Z_i\| = 1$, $Z_i \perp Z_j$ *for* $i \neq j$

(b) $\mathcal{L}(Z_1 \ldots Z_m) = \mathcal{L}(Y_1 \ldots Y_n)$

Proof. Use induction: suppose we have found $Z_1 \ldots Z_{n_k}$ satisfying (a) and

$$\mathcal{L}(Z_1 \ldots Z_{n_k}) = \mathcal{L}(Y_1 \ldots Y_k).$$

Now let $\hat{Y}_{k+1}$ be the orthogonal projection of Y_{k+1} onto $\mathcal{L}(Z_1 \ldots Z_{n_k})$, given by Proposition 1.1.1, and define

$$\tilde{Y}_{k+1} = Y_{k+1} - \hat{Y}_{k+1}$$

If $\|\tilde{Y}_{k+1}\| = 0$ then $Y_{k+1} \in \mathcal{L}(Z_1 \ldots Z_{n_k})$, so set $n_{k+1} = n_k$. Otherwise, set $n_{k+1} = n_k + 1$ and

$$Z_{n_{k}+1} = \frac{1}{\| \tilde{Y}_{k+1} \|} \tilde{Y}_{k+1}$$

If $U \in \mathcal{L}(Y_1 \ldots Y_{k+1})$ then there exists $\beta \in R^{k+1}$ such that

$$U = \sum_{i=1}^{k} \beta_i Y_i + \beta_{k+1} (\hat{Y}_{k+1} + \tilde{Y}_{k+1})$$

$$= \sum_{i=1}^{n_k} \gamma_i Z_i + \beta_{k+1} \| \tilde{Y}_{k+1} \| Z_{n_{k}+1}$$

for some $\gamma \in R^{n_k}$ since $\hat{Y}_{k+1} \in \mathcal{L}(Z_1 \ldots Z_{n_k})$. Thus (a) and (b) hold for $m = n_{k+1}, n = k + 1$. But for $k = 1$ we can define $n_k = 1$ and

$$Z_1 = \frac{1}{\| Y_1 \|} Y_1$$

Thus (a) and (b) hold with $m = n_n$.

The set $(Z_1 \ldots Z_m)$ is an *orthonormal (o.n.) basis* for $\mathcal{L}(Y_1 \ldots Y_n)$. Using Proposition 1.1.2 it is clear that Proposition 1.1.1 holds without the orthogonality of the Y_i, since these can be replaced by the orthonormal set $Z_1 \ldots Z_m$ which spans the same subspace. Of course the values of the α_i's calculated in (1.2) do depend on the orthogonality assumption.

The ideas of orthogonality, projection, etc., introduced above are related to the corresponding notions in Euclidean space R^n as follows.

1.1.3. Proposition. *Let* $Y_1 \ldots Y_n$ *be zero-mean finite-variance r.v.'s and let* $\mathcal{H} = \{ \sum_{i=1}^{n} \beta_i Y_i | \beta \in R^n \}$. *Then there is a one-to-one inner-product preserving map from* $\mathcal{H}$ *to Euclidean space* R^m *for some* $m \leqslant n$.

Proof. Let $Z_1 \ldots Z_m$ be an o.n. basis for $\mathcal{H}$. Then any $U \in \mathcal{H}$ has the unique representation

$$U = \sum_{i=1}^{m} \gamma_i Z_i$$

Define $\phi : \mathcal{H} \to R^m$ by

$$\phi(U) = \gamma = (\gamma_1 \dots \gamma_m)$$

This is clearly one-to-one and if $V = \Sigma \beta_i Z_i$ then

$$(U, V) = E\left(\sum \beta_i Z_i\right)\left(\sum \gamma_j Z_j\right) = \sum_i \beta_i \gamma_i = \beta \cdot \gamma$$

The inner-product preservation means in particular that both ϕ and ϕ^{-1} are continuous, i.e. ϕ is a homeomorphism. Thus $\mathcal{H}$ and R^m are geometrically identical, and subspaces, projections, etc., in $\mathcal{H}$ are exactly equivalent to the corresponding operations in R^m.

An immediate corollary of Proposition 1.1.3 is that the number m in Proposition 1.1.2 cannot depend on the ordering of the Y_i's since if a different ordering gave another o.n. basis $\tilde{Z}_1 \dots \tilde{Z}_{\tilde{m}}$ with, say $\tilde{m} > m$ then $\phi(\tilde{Z}_1) \dots \phi(\tilde{Z}_{\tilde{m}})$ would be a set of $\tilde{m}$ mutually orthogonal vectors in R^m.

Recursive estimation

In many practical cases the observed random variables $Y_1 \dots Y_n$ will represent measurements taken sequentially in time. Let $\hat{X}_k$ be the estimate of a related r.v. X given $Y_1 \dots Y_k$, i.e. the projection of X onto $\mathcal{L}(Y_1 \dots Y_k)$. Estimation is said to be *recursive* if $\hat{X}_k$ is obtained by 'updating' $\hat{X}_{k-1}$, that is, if $\hat{X}_k$ can be expressed in the form

$$\hat{X}_k = f_k(\hat{X}_{k-1}, Y_k). \tag{1.3}$$

If such functions f_k exist, (1.3) represents a computationally efficient procedure since it is then unnecessary to store all the past observations: only the current estimate and next observation are required. Using the geometric framework, we can easily see how recursive linear estimation can be accomplished. Let

$$L_k = \mathcal{L}(Y_1 \dots Y_k).$$

Then since $L_{k-1} \subset L_k$ we can always write

$$\hat{X}_k = A_k + B_k \tag{1.4}$$

where $A_k \in L_{k-1}$ and $B_k \perp L_{k-1}$. If $\tilde{X}_k = X - \hat{X}_k$ then

$$X = \hat{X}_k + \tilde{X}_k = A_k + (B_k + \tilde{X}_k)$$

and $A_k \in L_{k-1}$, $(B_k + \tilde{X}_k) \perp L_{k-1}$. But there is only one such decomposition of X, and it is given by

$$X = \hat{X}_{k-1} + \tilde{X}_{k-1}.$$

Thus $A_k = \hat{X}_{k-1}$. Now B_k is the projection of X onto $L_k -$ L_{k-1} (the orthogonal complement of L_{k-1} in L_k). If $L_k = L_{k-1}$ then $L_k - L_{k-1} = \{0\}$ so that $B_k = 0$. Otherwise this subspace is one-dimensional; indeed, denoting by $\mathscr{P}_k$ the projection onto L_k, we have

$$L_k - L_{k-1} = \mathcal{L}\{\tilde{Y}_k\} \tag{1.5}$$

where $\tilde{Y}_k = Y_k - \mathscr{P}_{k-1} Y_k$, obtained by the Gram–Schmidt orthogonalization procedure described in Proposition 1.1.2. The mutually orthogonal r.v.'s $\{\tilde{Y}_k, k = 1, 2, \ldots\}$ are called the *innovations sequence* and B_k is the *innovation projection* of X. From (1.2) and (1.5) this is given by

$$B_k = \frac{(X, \tilde{Y}_k)}{(\tilde{Y}_k, \tilde{Y}_k)} \tilde{Y}_k.$$

Thus (1.4) becomes

$$\hat{X}_k = \hat{X}_{k-1} + \frac{(X, \tilde{Y}_k)}{(\tilde{Y}_k, \tilde{Y}_k)} (Y_k - \mathscr{P}_{k-1} Y_k) \tag{1.6}$$

The name 'innovations' reflects the idea that the projection $\tilde{Y}_k$ represents the 'new information' acquired at time k, in the sense that the updating of $\hat{X}_{k-1}$ just depends on $\tilde{Y}_k$. Whether (1.6) can be turned into a recursive formula depends on the relation between $\mathscr{P}_{k-1} Y_k$ and $\hat{X}_{k-1}$. One case in which it can is given by the following example.

1.1.4. Example. Suppose $X, Z_1, Z_2 \ldots$ are independent zero-mean r.v.'s with

$$EX^2 = \sigma^2, \qquad EZ_i^2 = a^2$$

and

$$Y_k = X + Z_k.$$

Then $\hat{X}_k$ is given by the recursive formula

$$\hat{X}_k = \hat{X}_{k-1} + \frac{P_{k-1}}{P_{k-1} + a^2} (Y_k - \hat{X}_{k-1}), \hat{X}_0 = 0 \quad (1.7)$$

where $P_k = \| X - \hat{X}_k \|^2$ and is itself given recursively by

$$P_k = \frac{a^2 P_{k-1}}{a^2 + P_{k-1}} ; \quad P_0 = \sigma^2. \quad (1.8)$$

The point here is that $\mathscr{P}_{k-1} Z_k = 0$, so that

$$\mathscr{P}_{k-1} Y_k = \mathscr{P}_{k-1} X = \hat{X}_{k-1}.$$

Thus

$$\tilde{Y}_k = (X + Z_k) - \hat{X}_{k-1} = \tilde{X}_{k-1} + Z_k$$

and hence

$$(X, \tilde{Y}_k) = (X, \tilde{X}_{k-1}) = (\tilde{X}_{k-1}, \tilde{X}_{k-1}) = P_{k-1}$$
$$(\tilde{Y}_k, \tilde{Y}_k) = (\tilde{X}_{k-1} + Z_k, \tilde{X}_{k-1} + Z_k) = P_{k-1} + a^2$$

(1.7) follows by inserting these values in (1.6).

The estimate $\hat{X}_k = \mathscr{P}_k X$ can be computed directly as follows. Consider the sample mean

$$\bar{Y}_k = \frac{1}{k} \sum_{i=1}^{k} Y_i$$

and compute

$$(\bar{Y}_k, Y_i) = \frac{1}{k} \sum_{j=1}^{k} E(X + Z_i)(X + Z_j)$$

$$= \sigma^2 + \frac{1}{k} a^2$$

and

$$(X, Y_i) = \sigma^2$$

If we define

$$\alpha_k = \frac{\sigma^2}{\sigma^2 + \frac{1}{k} a^2}$$

we will then have

$$(X - \alpha_k \bar{Y}_k, Y_i) = 0$$

for all i; but this means $(X - \alpha_k \bar{Y}_k) \perp L_k$. Thus

$$\hat{X}_k = \alpha_k \bar{Y}_k \qquad (1.9)$$

Noting that we can write

$$\bar{Y}_k = \frac{1}{k}(Y_1 + \ldots + Y_k)$$

$$= \frac{k-1}{k} \bar{Y}_{k-1} + \frac{1}{k} Y_k,$$

we obtain

$$\hat{X}_k = \frac{k-1}{k} \frac{\alpha_k}{\alpha_{k-1}} \alpha_{k-1} \bar{Y}_{k-1} + \frac{\alpha_k}{k} Y_k$$

$$= \frac{k-1}{k} \frac{\alpha_k}{\alpha_{k-1}} \hat{X}_{k-1} + \frac{\alpha_k}{k} Y_k$$

$$= \hat{X}_{k-1} + \frac{\alpha_k}{k}(Y_k - \hat{X}_{k-1}) \qquad (1.10)$$

where the last line follows from the fact that

$$1 - \frac{\alpha_k}{k} = \frac{k-1}{k} \frac{\alpha_k}{\alpha_{k-1}}. \qquad (1.11)$$

Comparing (1.7) and (1.10) we see that

$$\frac{P_{k-1}}{P_{k-1} + a^2} = \frac{\alpha_k}{k}.$$

Hence, using (1.11)

$$P_{k-1} = \frac{a^2 \alpha_k}{k - \alpha_k} = \frac{a^2 \alpha_k \alpha_{k-1}}{(k-1)\alpha_k}$$

$$= a^2 \frac{\alpha_{k-1}}{k-1}$$

$$= \frac{a^2 P_{k-2}}{P_{k-2} + a^2}.$$

This is (1.8).

In this example the direct formula (1.9) is actually rather more enlightening than the recursion (1.7). $Y_1, Y_2, \ldots$ is just a sequence of 'measurements' of X with independent 'errors' $\{Z_i\}$. For large k, $\hat{X}_k \approx \bar{Y}_k$, which is the sample mean, whereas for small k the estimator takes account of the relative magnitudes of σ^2 and a^2: if the measurement error is large ($a^2 \gg \sigma^2$) then it more or less discards the observations and chooses a value close to the mean ($= 0$) of X.

Example 1.1.4 is a special case of the 'discrete-time Kalman filter'. One of the main concerns in Chapter 4 will be to generalize this to the situation where measurements are taken continuously in time and not just at a discrete set of time instants.

The Normal Case

A random n-vector W has the normal distribution $N(m, Q)$ if its characteristic function is the quadratic form

$$\phi_W(u) = Ee^{iu'W} = \exp(im'u - \tfrac{1}{2}u'Qu) \qquad (1.12)$$

where $m \in R^n$ and Q is a non-negative definite matrix. Then $m = EW$ and

$$Q = \mathrm{cov}(W) = E[(W - m)(W - m)']$$

It is immediate from (1.12) that
   (a) Linear combinations of normal r.v.'s are normal.
   (b) Every non-negative definite matrix Q is the covariance
       of a normal random vector. (Take $m = 0$ and define
       the characteristic function by (1.12).)
Recall that in the linear estimation problem of estimating X from $Y_1 \ldots Y_n$, only the means ($= 0$) and covariances were used. Thus it is no restriction to assume these r.v.'s are normally distributed since if not we can replace them by normal $\bar{X}, \bar{Y} \ldots \bar{Y}_n$ with the same covariance matrix. But in the normal case we get the stronger result that $\hat{X}$ is the best

non-linear estimator as well. Indeed, let $\tilde{X} = X - \hat{X}$. Then $\hat{X}, \tilde{X}$ are orthogonal, i.e. uncorrelated, and normal by (a) above. So they are independent. Compute the conditional characteristic function of X given $Y = (Y_1 \ldots Y_n)$:

$$\phi_{X|Y}(u; Y) = E(e^{iuX} | Y)$$
$$= E(e^{iu(\hat{X} + \tilde{X})} | Y)$$
$$= e^{iu\hat{X}} E(e^{iu\tilde{X}} | Y)$$
$$= e^{iu\hat{X}} E(e^{iu\tilde{X}}),$$

since $\hat{X} \in \mathcal{L}(Y)$ and $\tilde{X}$ is independent of Y.

Now $\tilde{X}$ has distribution $N(0, \sigma^2)$ where $\sigma^2 = \text{var}(\tilde{X})$, so that

$$\phi_{X|Y}(u; Y) = \exp(iu\hat{X} - \tfrac{1}{2}u^2\sigma^2)$$

But this is the characteristic function of $N(\hat{X}, \sigma^2)$. Thus $\hat{X}$ is the *conditional mean* of X:

$$\hat{X} = E(X|Y)$$

Recall the minimizing property of the expectation:

$$E(X - EX)^2 = \min_{\alpha \in R} E(X - \alpha)^2 \qquad (1.13)$$

Now consider the non-linear least-squares problem of finding the function $f(Y)$ that minimizes $E(X - f(Y))^2$. Let $g_{X|Y}(x; y)$ be the conditional density function of X given Y and $g_y(y)$ be the marginal density function of Y. Then

$$E(X - f(Y))^2 = \int_{R^n} g_y(y) \left(\int_R (x - f(y))^2 g_{X|Y}(x; y)dx \right) dy$$

Since $g_y(y) > 0$ for all y, this is minimized by minimizing the integrand pointwise for each y. But according to (1.13) this is achieved by

$$f(y) = E(X | Y = y) = \int_R x\, g_{X|Y}(x; y)\, dx = \hat{X}.$$

Thus $\hat{X}$, previously known to be the best *linear* estimator, is actually best in the class of all estimators, linear or not, because it is also the conditional mean.

To summarize: given the covariance of (Y, X) we can

calculate the best linear estimator $\hat{X}$ by projection. In general this will not be equal to the best non-linear estimator, which is $E(X|Y)$, but if in addition (Y, X) are normal then the two coincide.

This shows that the normal distribution has the curious property of being 'most random' in the sense that, of all possible distributions for (Y, X) having given a covariance matrix, the normal distribution maximizes the minimum estimation error. This is so because for any distribution

$$E(X - E(X|Y))^2 \leqslant E(X - \hat{X})^2$$

with equality for the normal distribution, the quantity on the right depending only on the covariance matrix.

1.2. Problems and Complements

1. Equation (1.1) is the general linear estimator of X given $Y_1 \ldots Y_n$. The general *affine* estimator is

$$\hat{X} = \sum_i \alpha_i Y_i + \beta.$$

Show that if X has mean m_X and the Y_i are independent with means $m_i = EY_i$, then the optimal affine estimator is

$$\hat{X} = \sum_i \alpha_i(Y_i - m_i) + m_X$$

with α_i given by (1.2). This shows that
  (a) Linear estimators are optimal in the zero-mean case.
  (b) The non-zero mean case can be reduced to this by
       considering the *centralized* r.v.'s $(Y_i - m_i)$, $(X - m_X)$.

2. The solution of (1.8) is

$$P_k = \frac{1}{\dfrac{k}{a^2} + \dfrac{1}{\sigma^2}} \tag{1.14}$$

This result can also be calculated directly from (1.9).

3. The variance equation (1.8) can be obtained from (1.7) by

writing (1.7) as

$$X - \hat{X}_k = (X - \hat{X}_{k-1}) - \frac{P_{k-1}}{P_{k-1} + a^2} \tilde{X}_k$$

and then computing the variance of the right hand side.

4. Suppose the r.v.'s in Example 1.1.4 are normal. Show by using Bayes' rule that the conditional distribution of X given $Y_1 \ldots Y_k$ is $N(\hat{X}_k, P_k)$ where $\hat{X}_k, P_k$ are given by (1.9) and (1.14). (The results above on the normal distribution imply that $\hat{X}_k$ is then the best linear estimator for arbitrary distributions, providing yet another way of solving the example.)

5. Suppose in Example 1.1.4, that instead of being constant the signal to be estimated is a *random walk*:

$$X_{k+1} = X_k + W_k, X_0 = X$$

where $\{W_k\}$ is a sequence of independent r.v.'s with zero mean and variance q^2, independent of $\{Z_k\}$. Denoting $\hat{X}_k = \mathscr{P}_k X_{k+1}$ and $P_k = \|X_{k+1} - \hat{X}_k\|^2$, show that $\hat{X}_k$ is still generated by (1.7) but that (1.8) is modified to

$$P_k = \frac{a^2 P_{k-1}}{a^2 + P_{k-1}} + q^2, \qquad P_0 = \sigma^2.$$

(Use the procedure suggested in Problem 3.)

CHAPTER TWO

Stochastic processes and linear estimation

In Chapter 1 we saw that any finite collection of r.v.'s could be regarded as elements of Euclidean n-space under the inner product

$$(X, Y) = EXY$$

and this is appropriate for minimum mean-square error linear estimation since the projection $\hat{Y}$ of an element Y onto a subspace $\mathcal{L}$ is closest to Y in the sense that

$$E(Y - \hat{Y})^2 = \| Y - \hat{Y} \|^2 \leqslant \| Y - Z \|^2 = E(Y - Z)^2$$

for any $Z \in \mathcal{L}$.

The programme for this chapter is to show that these ideas can be carried over to situations involving a possibly infinite number of r.v.'s. In Section 2.1 we introduce the probabilistic structure of some classes of stochastic processes, and in Section 2.2 the idea of Hilbert spaces, a class of linear spaces having many of the properties of Euclidean space but not being limited to finite dimensions. Finally in Section 2.3 we show that the relationship between stochastic processes and Hilbert space is exactly similar to that explored in Chapter 1 between random vectors and Euclidean space.

Bibliographic note. Cramér and Leadbetter's book [2] contains a beautifully clear introduction to stochastic process theory which includes among other things much of the material of this and the following chapter. An alternative account oriented towards dynamical systems is Wong [10]. Proofs of the fundamental results in measure theory stated but not proved below will all be found in Kingman and Taylor's text [4]. Standard reference works on probability and stochastic processes are Loève [7] and Doob [3].

2.1. Stochastic processes

Stochastic processes are mathematical models for random phenomena involving measurements at more than just a finite number of points. While more general interpretations are possible, we shall be concerned exclusively with situations where measurements are taken sequentially in time. Examples would be the 'noise' in electronic equipment, the random fluctuations of various quantities in a process control system, the movement of prices on the Stock Exchange. None of these cases can be represented realistically by a completely deterministic mathematical model, since this would imply, for example, the ability to predict exactly their future evolution. Thus we are forced back to probabilistic modelling which will enable us to say at least how the processes will behave in some average sense. The modelling question is not considered directly in this book, which is only concerned with properties of mathematical models, but let us remark that there are three main approaches:

(a) *Physical modelling* – based on direct analysis of the physical processes generating the fluctuations. Frequently used in electronics and communications, for example the analysis of 'shot noise' in vacuum tubes, or of fading, scattering, etc., in electromagnetic propagation.

(b) *'Black box' approach* – alias 'system identification'. Here no attempt is made to understand the physical nature of the processes. Instead a parametrized class of mathematical models is chosen and then parameters selected so that the model represents the available data in some 'best' way. Much used in process control.

(c) *Characterization results* – there are a number of results in mathematical statistics by which remarkably precise information about the distributions of random variables is implied by a small number of basic postulates. Thus often if a few simple conditions are met, the distributions can be inferred without further experimentation. The best-known such result is of course the central limit theorem which enables one to conclude that fluctuations arising as the superposition of a large number of independent random effects will be approximately normally distributed, whatever the distribution of the individual components.

Mathematical models of stochastic processes can be of varying degrees of complexity. Perhaps the simplest is the 'i.i.d.' sequence: a sequence of independent, identically distributed random variables $X_1, X_2, \ldots$. These can be used to represent errors in successive measurements of a quantity where there is no reason to believe either that the errors interact with each other or that they vary systematically in time. An i.i.d. sequence can also be used to build up more complicated processes, for example the ARMA (autoregressive moving average) process $Y_1, Y_2, \ldots$ obtained by solving the equations

$$Y_k + a_1 Y_{k-1} + \ldots + a_n Y_{k-n} = c_0 X_k + \ldots + c_n X_{k-n},$$

with given initial conditions. The behaviour of the (Y_k) sequence is obviously more complicated than that of the i.i.d. sequence X_k, and a variety of effects can be modelled by suitable choice of the coefficients (a_i, c_i). This is the type of model favoured in system identification.

The above are examples of *discrete-time* processes. In many cases a *continuous-time* model will be more appropriate. As a simple example, let X, ϕ be two random variables and define

$$Y_t = X \sin(\omega t + \phi);$$

this then represents a sinusoidal wave with random amplitude and phase. Thus for each t, Y_t is a random variable, in this case of a fairly simple type. For a more complicated example, consider the Poisson process, which is constructed as follows.

2.1.1. Definition. *Let* $S_1, S_2, \ldots$ *be an i.i.d. sequence of exponentially distributed r.v.'s with parameter* λ, *i.e.*

$$P[S_i > t] = e^{-\lambda t}, \qquad t \geqslant 0.$$

Let $T_0 = 0$ *and for* $n = 1, 2, \ldots$ *define*

$$T_n = \sum_{i=1}^{n} S_i.$$

Then the process N_t *defined by*

$$N_t = n \quad \text{if} \quad T_n \leqslant t < T_{n+1} \qquad n = 0, 1, 2, \ldots$$

is a Poisson process with rate λ.

According to this definition, N_t is simply the number of T_n's which are less than or equal to t, i.e. have occurred at or before t. Thus N_t is a r.v. which takes on integral values $0, 1, 2, \ldots$. One can show that it has the Poisson distribution and that $EN_t = \lambda t$, so that the 'rate' λ is the expected number of events per unit time. S_1, S_2, etc., are the *interarrival times* of the process. We return to the Poisson process in Section 3.2. It is frequently used as a model for 'rare' random events such as traffic flow on the uncongested road or particle emissions from a radioactive body.

Although the above examples were constructed in various ways, each gives rise to a random variable for each value t of the time set, related statistically to the r.v.'s at other times. We thus have the following definition.

2.1.2. Definition. *A* stochastic process *is a collection of random variables* $\{X_t : t \in T\}$ *indexed by a time set* $T \subset R$.

In this book we deal almost exclusively with *continuous-time* processes, for which the time set T will be either $R^+ = [0, \infty)$ or some finite interval $[a, b]$. For *discrete-time* processes T would be some subset of the integers $Z = \{0, \pm 1, \pm 2, \ldots\}$, often $Z^+ = \{0, 1, 2, \ldots\}$.

To proceed further it is necessary to introduce the axiomatic framework for probability and random variables. Only a sketch of this is given here and the reader is referred to [4, 5, 7, 10] for a full treatment. We introduce the *basic space* Ω whose elements ω are the *elementary events*, i.e. Ω is a list of all possible outcomes of some basic random experiment. The likelihood of occurrence of the various outcomes is evaluated by a probability measure P. If Ω is a finite or countable set, $\Omega = \{\omega_1, \omega_2, \omega_3, \dots\}$, then P is defined by giving, for each i, the probability p_i of ω_i. These will then satisfy:

$$p_i \geqslant 0, \sum_i p_i = 1. \qquad (2.1)$$

If Ω is bigger, then things are more complicated. For example (to jump ahead a bit) if X is a normal r.v. then the probability that X is *exactly* equal to any constant c is zero. But we can calculate the probability that, for example, $a \leqslant X \leqslant b$, i.e. of the interval $[a, b]$. Thus in the general framework we will have a set of *events* $\mathscr{B}$, each $B \in \mathscr{B}$ being a subset of Ω, i.e. a collection of elementary events, and P will be a set function, attaching a probability PB to each $B \in \mathscr{B}$. This assignment of probabilities must satisfy conditions analogous to (2.1), namely

$$0 \leqslant PB \leqslant 1$$
$$P\Omega = 1 \qquad (2.2)$$

$$\sum_i PB_i = P(\cup_i B_i) \qquad \text{for disjoint events } B_1, B_2, \dots$$

A *random variable* X is a real valued function $X : \Omega \to R$ with the property that, for each $a \in R$,

$$\{\omega : X(\omega) < a\} \in \mathscr{B}. \qquad (2.3)$$

The point of this is that probability is defined for the events B so that we can compute

$$F_X(a) = P\{\omega : X(\omega) < a\}$$

This is the *distribution function* (d.f.) of the random variable X. It is customary to suppress the ω-dependence of r.v.'s and to write $(X < a)$ in place of $\{\omega : X(\omega) < a\}$; this is less precise but has more intuitive content.

A good example of the above framework is provided by the roulette table. Here the basic random experiment is obvious and its outcomes are the numbers 0–36 (and 00 in Nevada). The assignment of probability is, on symmetry grounds, $P(\omega_i) = 1/37$ (1/38 in Nevada) for each i unless there is reason to believe that some hanky-panky is going on. In this case all subsets of Ω are events, including such well-known ones as 'red', 'even', etc. If there are N gamblers and they have placed their bets, then their winnings $X_i(\omega)$, $i = 1, 2, \ldots, N$ are random variables in accordance with our definition. A major advantage of the axiomatic framework is that the 'probability space' $(\Omega, \mathscr{B}, P)$ does not depend on the number of gamblers at the table, and in general one can have an arbitrary number of r.v.'s defined on the same probability space, a fact of some importance in connection with stochastic processes. But it should perhaps be pointed out that the above example is not quite as simple as it seems. First we have only considered one turn of the wheel. To account for two turns, Ω would need 37^2 points — all possible combinations of the outcomes of the first and second turns. Second, we have implicitly assumed that the gamblers place their bets in some deterministic manner. If one of them decides between two strategies by tossing a coin then the outcomes of this random experiment must be included in Ω, which then doubles in size. It is irritating to have to keep on enlarging Ω in this way. In practice it is conceptually and mathematically simpler to fix once and for all on a probability space which is 'big enough' to encompass all the random phenomena arising in the problem. We show below some ways in which this can be done.

In terms of measure theory a probability space $(\Omega, \mathscr{B}, P)$ is simply a measurable space $(\Omega, \mathscr{B})$ together with a positive measure P of total mass 1. That is, Ω is an arbitrary set, $\mathscr{B}$ is a σ-field of subsets of Ω, and P is a set function on $\mathscr{B}$ satisfying condition (2.2). A random variable is then a measurable function $X : (\Omega, \mathscr{B}) \to (R, \mathscr{S})$, where $\mathscr{S}$ is the σ-field of Borel sets in R, which means that $X^{-1}(S) \in \mathscr{B}$ for all $S \in \mathscr{S}$. (This is actually implied by the apparently weaker condition (2.3).)

The measure-theoretic framework is necessary when we come to consider the basic operation on random variables, namely calculating the expectation or mean value. In elementary probability theory this is defined as follows. If Ω is countable and $p_i = P(\{\omega_i\})$ then

$$EX = \sum_i X(\omega_i)p_i. \tag{2.4}$$

If, on the other hand the random variable X has a *density function*, i.e. there is a function f_X such that

$$F_X(a) = \int_{-\infty}^{a} f_X(x)\,dx,$$

then

$$EX = \int_{-\infty}^{\infty} xf_X(x)\,dx \tag{2.5}$$

We can subsume both these cases in a single formula by defining

$$EX = \int_{-\infty}^{\infty} x\,dF_X(x) \tag{2.6}$$

the quantity on the right being the *Stieltjes integral* of x with respect to the increasing function F_X. Its formal definition is indicated below.

Integration in the measure space $(\Omega, \mathscr{B}, P)$ is, briefly, defined in the following way (see [4] for the full story). First suppose the r.v. X is simple, i.e. just takes on a finite number of values $x_1 \ldots x_N$. Let $A_i = \{\omega \colon X(\omega) = x_i\}$. Then $X = x_i$ with probability PA_i and in analogy with (2.4) we define

$$\int_{\Omega} X(\omega)\,dP(\omega) = \sum_i x_i PA_i$$

Now for any positive random variable X it is possible to find a sequence X_n of simple r.v.'s which increase to X, i.e. such that

$$X_n(\omega) \uparrow X(\omega) \quad \text{for each} \quad \omega \in \Omega.$$

We now define

$$\int_{\Omega} X \mathrm{d}P = \lim_{n} \int_{\Omega} X_n \, \mathrm{d}P.$$

The sequence on the right is monotone increasing, so the limit is always unambiguously defined, but may be $+\infty$. Finally, for an arbitrary r.v. X we can write

$$X(\omega) = X^+(\omega) - X^-(\omega).$$

where

$$X^+(\omega) = X(\omega)I_{(X \geqslant 0)}$$
$$X^-(\omega) = -X(\omega)I_{(X < 0)}$$

Here I_A, the *indicator function* of the set A, is defined by

$$I_A(\omega) = \begin{cases} 1 & \omega \in A \\ 0 & \omega \notin A \end{cases}.$$

If both $\int_{\Omega} X^+ \mathrm{d}P$ and $\int_{\Omega} X^- \mathrm{d}P$ are finite we define

$$\int_{\Omega} X \mathrm{d}P = \int_{\Omega} X^+ \mathrm{d}P - \int_{\Omega} X^- \mathrm{d}P.$$

Otherwise the integral is not defined. Note that $|X| = X^+ + X^-$ and this is positive, so that the condition that $\int_{\Omega} X \mathrm{d}P$ exists is just that $\int_{\Omega} |X| \, \mathrm{d}P < \infty$. X is said to be *integrable* if this condition holds.

Suppose X and Y are r.v.'s and let

$$D = \{\omega : X(\omega) \neq Y(\omega)\}.$$

Then D is an event, i.e. $D \in \mathscr{B}$. If $PD = 0$, X and Y are 'almost' the same in the sense that $X = Y$ with probability one. X and Y are then said to be *equivalent* r.v.'s and we write $X = Y$ a.s. (= almost surely). Similarly, if X_n is a sequence of r.v.'s then the statement '$X_n \to X$ a.s.' means: there is a set $F \in \mathscr{B}$ such that $PF = 0$ and $X_n(\omega) \to X(\omega)$ for each $\omega \notin F$. In this case if we define

$$\tilde{X}_n(\omega) = \begin{cases} X_n(\omega) & \omega \notin F \\ X(\omega) & \omega \in F \end{cases}$$

then $\tilde{X}_n = X_n$ a.s. and $\tilde{X}_n(\omega) \to X(\omega)$ for every ω, so that to say $X_n \to X$ a.s. is only slightly weaker than saying $X_n(\omega) \to X(\omega)$ pointwise.

The operation of integration does not distinguish between equivalent r.v.'s. In this regard, the properties of the integral that we shall need are summarized as follows.

2.1.3. Proposition. *(a) If X is an integrable r.v. and Y is an r.v. such that $X = Y$ a.s., then Y is integrable and*

$$\int_{\Omega} X \mathrm{d}P = \int_{\Omega} Y \mathrm{d}P$$

(b) (Bounded convergence theorem) If X_n and X are r.v.'s such that

(i) $X_n \to X$ a.s.

(ii) there exists a constant K such that, for each n,

$$|X_n| < K \text{ a.s.}$$

then

$$\int_{\Omega} X_n \mathrm{d}P \to \int_{\Omega} X \mathrm{d}P.$$

(c) If $X \geqslant 0$ a.s. then $\int_{\Omega} X \mathrm{d}P = 0$ implies $X = 0$ a.s.

Proof. See [4], Theorems 5.5 and 5.8. For (c), note that, directly from the definition, $\int_{\Omega} X \mathrm{d}P \geqslant (1/n) P A_n$ where $A_n = \{\omega : X(\omega) \geqslant 1/n\}$ so that $P A_n = 0$ for all n if $\int_{\Omega} X \mathrm{d}P = 0$. Thus $P\{\omega : X(\omega) > 0\} = P(\cup_n A_n) = 0$.

The significance of the distribution function F_X of a r.v. X is that it defines a measure μ_X on the *sample space* $(R, \mathscr{S})$, determined by the recipe

$$\mu_X([a, b)) = F_X(b) - F_X(a). \tag{2.7}$$

We now have a new probability space $(R, \mathscr{S}, \mu_X)$ and can define integrals of the form

$$\int_R g(x) \mathrm{d}\mu_X(x) \tag{2.8}$$

as before, for integrable functions $g : R \to R$. The integral is also denoted by

$$\int_{-\infty}^{\infty} g(x)\mathrm{d}F_X(x)$$

and is called the *Lebesgue–Stieltjes integral* of g with respect to the d.f. F_X. This is how the definition (2.6) of EX is to be interpreted. It can be shown that the value of EX obtained coincides with the values of (2.4) and (2.5) in those special cases.

The relation between the various concepts introduced above is as follows.

2.1.4. Proposition. *For any measurable function* $g : R \to R$,

$$\int_{-\infty}^{\infty} g(x)\mathrm{d}F_X(x) = \int_{\Omega} g\circ X(\omega)\mathrm{d}P(\omega) \qquad (2.9)$$

Here the left hand side is defined by (2.7)–(2.8) and the statement means that if either side exists then so does the other, and they are equal.

Proof. By definition, (2.7) says

$$\mu_X([a, b)) = P\{\omega : a \leqslant X(\omega) < b\}$$
$$= P[X^{-1}([a, b))].$$

In fact this property extends to the Borel sets $\mathscr{S}$, so that

$$\mu_X S = P(X^{-1}(S)) \quad \text{for all} \quad S \in \mathscr{S}.$$

First suppose g is a simple function, i.e.

$$g(x) = \sum_i g_i I_{S_i}(x).$$

Then by definition

$$\int g\mathrm{d}F_X = \sum_i g_i \mu_X S_i.$$

But then $g\circ X$ is also a simple function, taking the value g_i on the set $X^{-1}(S_i)$. Thus

$$\int_{\Omega} g\circ X(\omega)\mathrm{d}P(\omega) = \sum_i g_i P(X^{-1}(S_i))$$

so that (2.9) holds for simple g. For positive g, take a sequence of simple functions $g_n \uparrow g$. Then $g_n \circ X \uparrow g \circ X$ so that

$$\int_\Omega g \circ X \mathrm{d}P = \lim_n \int_\Omega g_n \circ X \mathrm{d}P = \lim_n \int_R g_n \mathrm{d}\mu_X = \int_{-\infty}^\infty g \mathrm{d}F_X(x).$$

Thus (2.9) holds for positive g and the extension to arbitrary g proceeds as before.

This result means in particular that

$$EX = \int_\Omega X(\omega) \mathrm{d}P(\omega).$$

Thus for any random variable on a fixed probability space $(\Omega, \mathscr{B}, P)$ the expectation is defined in terms of a single measure P. This is conceptually simpler than having to deal always with the sample space measure μ_X or distribution function F_X, which is different for each r.v.

Construction of Stochastic processes

According to Definition 2.1.2, a stochastic process is an indexed family of r.v.'s $\{X_t, t \in T\}$. These must be defined on a fixed probability space $(\Omega, \mathscr{B}, P)$. Notice that the process can then be regarded as a single function $X: T \times \Omega \rightarrow R$. Fixing t we get the r.v. $X_t = X(t, .\ )$, while fixing ω gives the *sample function* $X(.\ , \omega) = \{X(t, \omega): t \in T\}$, the values of each of the r.v.'s X_t for a particular elementary event, or sample, ω.

To specify a process probabilistically, an appropriate probability space $(\Omega, \mathscr{B}, P)$ to carry the r.v.'s must be supplied. In some cases, such as the roulette table, this arises naturally, but more commonly the distributions of the r.v.'s are the primary entities and the problem arises of constructing $(\Omega, \mathscr{B}, P, X_t)$ in such a way that the distributions of the $\{X_t\}$ have certain specified properties. For a single r.v. required to have distribution function F_X, this is no problem: we can take $(\Omega, \mathscr{B}) = (R, \mathscr{S})$, $P = \mu_X$ (defined by (2.7)) and, for the r.v. X the identity function

$$X(\omega) = \omega.$$

In this case the introduction of $(\Omega, \mathcal{B}, P)$ is rather super-fluous and it is easier to think in terms of the 'sample space' $(R, \mathcal{S}, \mu_X)$; but, as remarked above, where a lot of r.v.'s are involved the probability space idea offers a genuine simplification.

The general solution to the construction problem is actually the result quoted as Proposition 2.1.7 below. But that is a much more powerful result than is required for the purposes of this book. The clue to what we require is provided by the Poisson process of Definition 2.1.1, where the r.v.'s (N_t) are constructed from a basic sequence of i.i.d. interarrival times $S_1, S_2, \ldots$. Thus the probabilistic framework for the Poisson process is provided if we can specify a probability space carrying an i.i.d. sequence of random variables with a given distribution function F. In fact the same is true for *every* process encountered in this book. So the following result suffices for our purposes.

2.1.5. Proposition. *Let F be a distribution function. Then there exists a probability space $(\Omega, \mathcal{F}, P)$ and a stochastic process $\{X_k, k \in Z^+\}$ such that for any $k_1, k_2 \ldots k_n \in Z^+$ and $a_1 \ldots a_n \in R$*

$$P\{\omega: X_{k_1} < a_1, X_{k_2} < a_2 \ldots X_{k_n} < a_n\} = \prod_{i=1}^{n} F(a_i).$$
$$(2.10)$$

This says that (X_k) is an i.i.d. sequence with common d.f. F.

Proof*. For $k = 0, 1, 2, \ldots$, let

$$(\Omega_k, \mathcal{B}_k, \mu_k) = (R, \mathcal{S}, \mu)$$

where μ is the measure corresponding to F via (2.7). Now take

$$\Omega = \prod_{k=0}^{\infty} \Omega_k$$

and let $\mathcal{B}$ be the σ-field in Ω generated by the finite-dimensional cylinder sets, i.e. the smallest σ-field containing all sets A of the form

$$A = B_0 \times B_1 \times \ldots \times B_n \times \prod_{k=n+1}^{\infty} \Omega_k .$$

Then according to Theorem 6.3 of [4] there is a unique *product measure* P on $(\Omega, \mathscr{B})$ with the property that $PA = \mu(B_0)\mu(B_1) \ldots \mu(B_n)$. Let X_i be the ith *coordinate function*: $X_i(\omega) = \omega_i$, where $\omega = (\omega_0, \omega_1, \ldots)$. Then the X_i are independent and each has d.f. F.

Using the above construction the measure P depends on the d.f. F of the sequence. It is possible to go further and standardize the measure, so that one probability space suffices for all i.i.d. sequences whatever their distribution. A r.v. U is *uniform* $[0, 1]$ if it has the density function

$$f_U(x) = \begin{cases} 1 & 0 \leqslant x \leqslant 1 \\ 0 & \text{otherwise} \end{cases}$$

U has the property that $P(U < x) = x$ for $x \in [0, 1]$.

2.1.6. Proposition. *Let F be a distribution function and U be a uniform $[0, 1]$ r.v. Then there exists a function $g:[0, 1] \to R$ such that the r.v. $g(U)$ has d.f. F.*

Proof. If F is continuous and strictly increasing then the appropriate function is $g(u) = F^{-1}(u)$. For since F is monotone, $F^{-1}(u) < x \Leftrightarrow u < F(x)$ and hence

$$P(g(U) < x) = P(U < F(x)) = F(x). \qquad (2.11)$$

For general F the inverse may not be a well-defined function, but if we take

$$g(u) = \sup \{t : F(t) < u\} \qquad (2.12)$$

(which coincides with F^{-1} where this is defined) then it is easily checked that (2.11) still holds.

From Proposition 2.1.5 we know that there is a probability space $(\Omega, \mathscr{B}, P)$ and a process $\{U_k, k \in Z^+\}$ such that $\{U_k\}$ is an i.i.d. sequence of uniform $[0, 1]$ r.v.'s. If we now define

$$X_k(\omega) = g \circ U_k(\omega) = g(U_k(\omega))$$

then Proposition 2.1.6 shows that $\{X_k\}$ is an i.i.d. sequence carried by $(\Omega, \mathcal{B}, P)$ with d.f. F. This may seem at first sight a particularly academic piece of tidying-up, but it is in fact a result of great practical significance in connection with simulation, since a sequence of i.i.d. uniform $[0, 1]$ r.v.'s is what is normally produced by a standard library procedure for generating random numbers in a computer. (To be exact, what is produced is a sequence of numbers which cannot be distinguished from a sample function of $\{U_k\}$ by statistical tests.) Proposition 2.1.6 shows that any i.i.d. sequence can be generated from this, and hence that every process in this book can be simulated using standard library procedures. It should be remarked, however, that the method suggested by (2.12) may not be the most efficient way of doing this in particular cases, since it involves repeated evalution of the possibly quite complicated function g.

Parameters of Processes

Let $\{X_t, t \in T\}$ be a stochastic process. Suppose $E \, | X_t | < \infty$ for all $t \in T$. Then the function

$$m(t) = EX_t$$

is the *mean* of the process. It is a non-random function $m : T \to R$. The 'centred' process

$$\tilde{X}_t = X_t - m(t)$$

has zero mean and since this only differs from X_t by a deterministic quantity it is often convenient to consider the centred version of a process rather than the process itself. *Any* function could in principle be the mean function for some stochastic process.

Now suppose in addition that $EX_t^2 < \infty$ for all t. This actually implies the previous assumption $E \, | X_t | < \infty$ since, if F is the distribution function of X_t then

$$E \, | X_t | = \int\limits_{-\infty}^{\infty} | x | \, dF(x)$$

$$= \int_{[|x| \leqslant 1]} |x| \, dF(x) + \int_{[|x| > 1]} |x| \, dF(x)$$

$$\leqslant \int_{[|x| \leqslant 1]} dF(x) + \int_{[|x| > 1]} x^2 \, dF(x)$$

$$\leqslant 1 + EX_t^2 .$$

One can show [4, Theorem 7.7] that the *Schwarz inequality* also holds:

$$E|X_t X_s| \leqslant \sqrt{EX_t^2 EX_s^2}$$

Thus X_t and $X_t X_s$ are integrable and we can define the *covariance function*:

$$r(t, s) = E(X_t - m(t))(X_s - m(s)).$$

It is a symmetric function on $T \times T$ to R. It is also non-negative definite, in the sense that for any $t_1 \ldots t_n \in T$ and $\xi = (\xi_1, \xi_2 \ldots \xi_n) \in R^n$,

$$\sum_{ij} \xi_i \xi_j r(t_i, t_j) \geqslant 0.$$

This is because

$$0 \leqslant E \left(\sum_i \xi_i (X_{t_i} - m_{t_i}) \right)^2 = E \sum_{ij} \xi_i \xi_j (X_{t_i} - m_{t_i})(X_{t_j} - m_{t_j})$$

$$= \sum_{ij} \xi_i \xi_j r(t_i, t_j).$$

Other properties of covariance functions are given in Proposition 2.1.8 below.

The Kolmogorov consistency theorem and normal processes

It was shown in Chapter 1 that *every* non-negative definite matrix is the covariance matrix of a normal random vector. An analogous result holds for processes; but to show this we have to consider other probability spaces than those introduced in Proposition 2.1.5. Let $\{X_t, t \in T\}$ be a stochastic process. Then for any $t_1, t_2 \ldots t_n \in T, (X_{t_1}, X_{t_2} \ldots X_{t_n})$ is a random n-vector. Its distribution function is

$$F_{t_1 \ldots t_n}(a_1 \ldots a_n) = P(X_{t_1} < a_1, X_{t_2} < a_2 \ldots X_{t_n} < a_n).$$

$$(2.13)$$

The set of $(F_{t_1 \ldots t_n})$ for all possible choices of n and $t_1 \ldots t_n$ is called the *family of finite-dimensional distributions* of the process. They satisfy the following two obvious properties.

(i) For any permutation π of $(1, 2 \ldots n)$,

$$F_{t_{\pi(1)} \ldots t_{\pi(n)}}(a_{\pi(1)} \ldots a_{\pi(n)}) = F_{t_1 \ldots t_n}(a_1 \ldots a_n).$$

This just says, for example that $F_{t_1, t_2}(a_1, a_2) = F_{t_2, t_1}(a_2, a_1)$ which is clear since these are just alternative ways of writing $P(X_{t_1} < a_1, X_{t_2} < a_2)$

(ii)

$$\lim_{a_i \uparrow \infty} F_{t_1 \ldots t_n}(a_1 \ldots a_n) =$$

$$F_{t_1 \ldots t_{i-1}, t_{i+1}; \ldots t_n}(a_i \ldots a_{i-1}, a_{i+1} \ldots a_n).$$

The expression on the left is just $P(X_{t_1} < a_1, \ldots X_{t_i} < \infty, \ldots X_{t_n} < a_n)$, which places no restrictions on the ith coordinate. Properties (i) and (ii) are known as the Kolmogorov consistency conditions. It transpires that they are actually sufficient for a candidate family of distribution functions to correspond to some stochastic process. This result is due to Kolmogorov and is in a sense the fundamental theorem of stochastic processes.

2.1.7. Proposition. *Suppose that for each n and $t_1 \ldots t_n \in T$ a distribution function $F_{t_1 \ldots t_n}$ is given and that the family of these satisfies the consistency conditions (i) and (ii) above. Then there exists a probability space $(\Omega, \mathscr{F}, P)$ and a stochastic process $\{X_t, t \in T\}$ such that (2.13) holds, i.e. such that $(F_{t_1 \ldots t_n})$ is the family of finite dimensional distributions for the process $\{X_t\}$.*

Proof. See Theorem 6.10 of [4].

As remarked earlier, we do not need this result directly since all the processes we consider will be constructed from i.i.d. sequences for which Proposition 2.1.4 provides the framework. However we do need it to elucidate the relation between arbitrary and normal processes. A *normal* (or *gaussian*) process is one each of whose finite-dimensional distributions is normal. We showed in above that if $r(t, s)$ is the covariance function of a process $\{X_t\}$ then $r(t, s)$ is non-negative definite. Proposition 2.1.8 provides the converse.

2.1.8. Proposition. *Let $r(t, s)$, $t, s \in T$ be a non-negative definite function. Then there exists a* normal *process $\{X_t, t \in T\}$ such that*

$$r(t, s) = \text{cov}(X_t, X_s).$$

Proof. Choose $t_1 \ldots t_n \in T$. Then the matrix

$$Q = [r(t_i, t_j)] = \begin{bmatrix} r(t_1 t_1) & r(t_1 t_2) \ldots r(t_1 t_n) \\ \vdots & r(t_2 t_2) \quad \vdots \\ r(t_n t_1) & \quad r(t_n t_n) \end{bmatrix}$$

is non-negative definite by assumption, and hence

$$\phi_{X_1 \ldots X_n}(u) = e^{-\frac{1}{2} u' Q u} \tag{2.14}$$

is the characteristic function of a normal random vector $X_1 \ldots X_n$ with zero mean and covariance matrix Q. It is easily checked that the family of finite dimensional distributions defined by (2.14) satisfies the consistency conditions (i) and (ii). Hence by Proposition 2.1.7 there is a process $\{X_t, t \in T\}$ whose finite dimensional distributions are (2.14). Thus X_t is a normal process and from the two-dimensional distribution we see that

$$\text{cov}(X_t, X_s) = r(t, s).$$

This result shows, first, that there is a one-to-one relationship between non-negative definite functions and normal stochastic processes, and, secondly, that for every finite-variance process there is a normal process with the same covariance function. This corresponds exactly to the finite dimensional result that for every random vector X there is a normal random vector Y with the same covariance matrix.

2.2. Hilbert Space

A *real linear vector space* is a set X on which are defined addition and scalar multiplication operations which have the same properties as those operations in R^n. That is, there are functions $f: X \times X \to X$ and $g: R \times X \to X$ having the properties listed below. We write $f(x, y) = x + y$ and $g(c, x) = cx$. Then for all $x, y, z \in X$ and $a, b \in R$,

(i) $x + y = y + x$ (commutative law)

(ii) $(x + y) + z = x + (y + z)$ (associative law)

(iii) there exists $0 \in X$ such that $x + 0 = x$

(iv) $a(x + y) = ax + ay$

(v) $(a + b)x = ax + bx$ (distributive laws)

(vi) $abx = a(bx)$

(vii) $0x = 0, 1x = x$

Notice that 0 denotes the null elements of both R and X. This will cause no confusion. Also we write $x - y = x + (-1)y$.

A *subspace M* of X is a subset which is closed under linear operations, i.e. $x, y \in M$ implies $ax + by \in M$ for all $a, b \in R$. If $\{M_\alpha, \alpha \in I\}$ is a collection of subspaces indexed by any arbitrary set I then $\bigcap_{\alpha \in I} M_\alpha$ is also a subspace, as is easily checked. For any subset $Q \subset X$, the subspace *spanned* or *generated* by Q, denoted by $\mathcal{L}(Q)$, is the set of linear combinations of the elements of Q. $\mathcal{L}(Q)$ can be characterized as the smallest subspace containing Q, i.e. the intersection of all subspaces containing Q, since Q is contained in at least one subspace, namely X itself, and the intersection of all such subspaces is a subspace, as noted above. Note that 0 (the zero element of X) is in every subspace and that $\{0\}$ (the set consisting of the single element 0) is a subspace.

A *pre-Hilbert* space is a real linear vector space $\mathcal{H}$ on which is defined an *inner product* function $h: \mathcal{H} \times \mathcal{H} \to R$ satisfying the following properties (for all $x, y, z \in \mathcal{H}$ and $a, b \in R$). We write $h(x, y) = (x, y)$.

(i) $(y, x) = (x, y)$

(ii) $(x, x) \geqslant 0; (x, x) = 0 \Leftrightarrow x = 0$ (2.15)

(iii) $(ax + by, z) = a(x, z) + b(y, z)$.

The *norm* of $x \in \mathcal{H}$ is

$$\| x \| = \sqrt{(x, x)}.$$

2.2.1. Proposition. *For any* $x, y \in \mathcal{H}$ *we have*

    (a) the Schwarz Inequality: $|(x, y)| \leqslant \| x \| \cdot \| y \|$

    (b) the Triangle Inequality: $\| x + y \| \leqslant \| x \| + \| y \|$

    (c) the Parallelogram Equality:

$$\| x + y \|^2 + \| x - y \|^2 = 2 \| x \|^2 + 2 \| y \|^2$$

Proof. Using the axioms (2.15) we have that for any $a \in R$, $x, y \in \mathcal{H}$,

$$0 \leqslant (ax + y, ax + y) = a^2 (x, x) + 2a(x, y) + (y, y)$$

$$= (x, x) \left\{ \left(a + \frac{(x, y)}{(x, x)} \right)^2 + \frac{(y, y)}{(x, x)} - \frac{(x, y)^2}{(x, x)^2} \right\}$$

If $(x, x) = 0$ then $(x, y) = 0$ from (2.15) (ii) and (iii) (with $a = b = 0$). Otherwise take $a = -(x, y)/(x, x)$ to give the Schwarz inequality. Now taking $a = 1$,

$$\| x + y \|^2 = \| x \|^2 + 2(x, y) + \| y \|^2$$

$$\leqslant \| x \|^2 + 2 \| x \| \cdot \| y \| + \| y \|^2 \quad \text{(Schwarz)}$$

$$= (\| x \| + \| y \|)^2 \quad\quad\quad\quad (2.16)$$

Hence (b). Replacing y by $-y$ in (2.16) and adding gives (c).

    We can now use the inner product to define a topology, i.e. a notion of convergence, in $\mathcal{H}$. The *distance* between two points is defined as

$$d(x, y) = \| x - y \|.$$

This is a *metric* on $\mathcal{H}$, thanks to the axioms (2.15) and the triangle inequality, since it satisfies

$$d(x, y) = d(y, x)$$

$$d(x, y) = 0 \Leftrightarrow x = y \quad\quad\quad (2.17)$$

$$d(x, z) \leqslant d(x, y) + d(y, z)$$

and we say that a sequence (x_n) *converges* to x $(x_n \to x)$ if $d(x_n, x) \to 0$ as $n \to \infty$. All statements involving convergence in $\mathcal{H}$ are to be interpreted in this sense. The first result is the following.

2.2.2. Proposition. *The inner product is a continuous function from $\mathcal{H} \times \mathcal{H}$ to R, i.e. if $x_n \to x$ and $y_n \to y$ then (x_y, y_n) $\to (x, y)$. In particular, $x_n \to x$ implies $\| x_n \| \to \| x \|$.*

Proof. If x_n is a convergent sequence then x_n is bounded, i.e. $\| x_n \| < K$ for some $K < \infty$ (choose N such that $\| x_n - x \| < 1$ for $n \geqslant N$ and let $K = \| x \| + 1 + \max_{1 \leqslant i \leqslant N} \| x_i \|$). Now

$$(x, y) = ((x - x_n) + x_n, (y - y_n) + y_n)$$
$$= (x - x_n, y - y_n) + (x - x_n, y_n) + (x_n, y - y_n) + (x_n, y_n)$$

The first three terms on the right converge to zero, since, for example

$$|(x_n, y - y_n)| \leqslant \| x_n \| \cdot \| y - y_n \| \leqslant K \| y - y_n \|.$$

The result follows.

If x_n is a convergent sequence with limit x, then from the triangle inequality

$$d(x_n, x_m) \leqslant d(x_n, x) + d(x_m, x)$$

for any n, m, so that $d(x_n, x_m) \to 0$ as $n, m \to \infty$. Any sequence in $\mathcal{H}$ having this property is called a *Cauchy sequence*. $\mathcal{H}$ is said to be *complete* if the converse holds, i.e. if for every Cauchy sequence (x_n) there exists some $x \in \mathcal{H}$ such that $x_n \to x$.

2.2.3. Definition. *A* Hilbert space *is a pre-Hilbert space which is complete with respect to the metric* $d(x, y) = \| x - y \|$.

To add content to this definition, consider the real line R. This is a Hilbert space under the usual algebraic operations and the inner product $(x, y) = xy$. The rational numbers Q form a pre-Hilbert space under the same operations but this is not complete since, for example

$$x_n = 1 + 1 + \frac{1}{2!} + \ldots + \frac{1}{n!}$$

is a Cauchy sequence in Q which does not converge to any *rational* number. Thus completeness expresses the idea that there are no 'gaps'. Let us consider some other spaces.

1. R^n *(Euclidean space)*. Vector addition, scalar multiplication and

$$(x, y) = \sum_{i=1}^{n} x_i y_i$$

Its completeness is an immediate consequence of that of R.

The interest in Hilbert space lies in the fact that the concept extends beyond finite-dimensional spaces. Usually, verifying that a candidate space is pre-Hilbert is no problem, the delicate part being to show that it is complete.

2. $C[0, 1]$ (the space of continuous functions $f: [0, 1] \to R$.) Here we define

$$(f + g)(t) = f(t) + g(t)$$

$$(cf)(t) = cf(t)$$

$$(f, g) = \int_{0}^{1} f(t) g(t) \, dt$$

These satisfy all the axioms, so $C[0, 1]$ is a pre-Hilbert space; however, it is *not* complete. Consider, for example, the sequence

$$f_n(t) = \begin{cases} (2t)^n & 0 \leqslant t \leqslant \frac{1}{2} \\ 1 & \frac{1}{2} \leqslant t \leqslant 1 \end{cases}$$

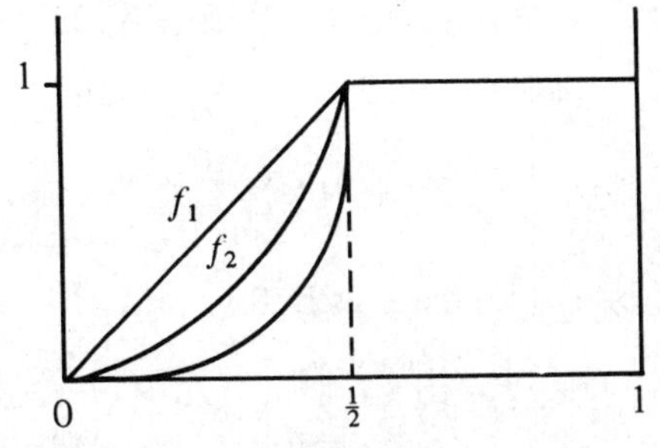

Fig. 2.1

(see Fig. 2.1). This is a Cauchy sequence since for $m > n$

$$\|f_n - f_m\|^2 = \int_0^1 (f_n(t) - f_m(t))^2 \, dt$$

$$\leqslant \int_0^{\frac{1}{2}} (f_n(t))^2 \, dt \to 0 \quad \text{as} \quad n \to \infty$$

But there cannot be any *continuous* function such that $\|f_n - f\| \to 0$. This means that $C[0, 1]$ is 'not big enough'. A space which is, is:

3. $L_2[0, 1] = \{f : [0, 1] \to R \mid \int_0^1 f^2(t) \, dt < \infty\}$ with the same operations as before. In fact, $L_2[0, 1]$ is exactly the completion of $C[0, 1]$ with respect to the metric

$$d(f, g) = \int_0^1 (f(t) - g(t))^2 \, dt. \tag{2.18}$$

This means that every $f \in L_2$ is the limit of a Cauchy sequence of functions which are continuous.

The proof of these facts follows from a general result in measure theory given as Theorem 2.3.1 below. An important remark is that they depend on the integrals involved being interpreted in the sense of Lebesgue integration. $L_2[0, 1]$ is *not* complete with respect to, say, Riemann integration, since it is possible for a sequence of Riemann-integrable functions to converge in the metric (2.18) to a function which is not Riemann integrable.

A final example:

l_2 *(square summable sequences)*. Formally, if

$$Z^+ = \{0, 1, 2 \ldots\} \quad \text{then}$$

$$l_2 = \left\{ x : Z^+ \to R \mid \sum_{i=0}^{\infty} (x(i))^2 < \infty \right\}$$

The algebraic operations are as before and

$$(x, y) = \sum_{i=0}^{\infty} x(i) y(i)$$

The completeness is another corollary of Theorem 2.3.1. This space is mainly of use in the analysis of discrete-time systems, which are not considered in this book.

The most important property of Hilbert spaces is that ideas of orthogonality and projection apply in them just as in finite-dimensional spaces.

Subspaces of linear vector spaces have been defined above. In a Hilbert space $\mathcal{H}$, we only consider *closed* subspaces, i.e. those containing all limits of sequences of their elements. (The reason for this will become apparent in the proof of Theorem 2.2.4 below.) Thus $M \subset \mathcal{H}$ is a subspace if

(i) $x, y \in M, a, b \in R \Rightarrow (ax + by) \in M$

(ii) (x_n) is a Cauchy sequence in M

$$\Rightarrow x = \lim x_n \in M$$

Note that subspaces of R^n are automatically closed, but this is not true of every Hilbert space.

Two elements $x, y \in \mathcal{H}$ are *orthogonal* $(x \perp y)$ if $(x, y) = 0$. If $Y \subset \mathcal{H}$ is any subset then $x \perp Y$ means $x \perp y$ for all $y \in Y$. In accordance with our convention above $\mathcal{L}(Y)$ denotes the smallest *closed* subspace containing Y. Note that $x \perp Y$ implies $x \perp \mathcal{L}(Y)$ by Proposition 2.2.2. Finally if M is a subspace we define

$$M^\perp = \{v \in \mathcal{H} : v \perp M\}$$

By Proposition 2.2.2, this is again a subspace.

Here now is the main result.

2.2.4. Theorem. *Let $\mathcal{H}$ be a Hilbert space, and $M \subset \mathcal{H}$ be a (closed) subspace. Then any element $x \in \mathcal{H}$ has the unique decomposition*

$$x = y + z$$

where

$$y \in M, \quad z \perp M.$$

Furthermore

$$\|x - y\| = \min_{v \in M} \|x - v\|$$

Proof. To take the uniqueness first, suppose two pairs of

elements x, y and x', y' have the asserted properties. Then
$x = y + z = y' + z'$ so that

$$y - y' = z' - z.$$

But $(y - y') \in M$ and $(z' - z) \in M^\perp$. Hence $(y - y', y - y')$
$= (y - y', z - z') = 0$ so that $\|y - y'\| = 0$; hence $y = y'$,
$z = z'$. Thus at most one pair x, y can satisfy the requirements. If $x \in M$ then $x = x + 0$ is the asserted decomposition, so assume $x \notin M$, in which case

$$\inf_{v \in M} \|x - v\| = h > 0 \qquad (2.19)$$

since M is closed. Let y_n be a sequence in M such that
$\|y_n - x\| \downarrow h$. The parallelogram equality states that for
$u, v \in \mathcal{H}$

$$\|u + v\|^2 + \|u - v\|^2 = 2\|u\|^2 + 2\|v\|^2$$

Taking $u = y_n - x, v = y_m - x$ gives

$$\|y_n + y_m - 2x\|^2 + \|y_n - y_m\|^2 = 2\|y_n - x\|^2 + 2\|y_m - x\|^2$$

Now $\hspace{6cm} (2.20)$

$$\|y_n + y_m - 2x\|^2 = 4\|\tfrac{1}{2}(y_n + y_m) - x\|^2 \geqslant 4h^2$$

since $\tfrac{1}{2}(y_n + y_m) \in M$. Fix $\epsilon > 0$ and choose N such that
for all $n, m > N$

$$\|y_n - x\|^2 < \tfrac{1}{4}\epsilon + h^2$$

Then from (2.20)

$$\|y_n - y_m\|^2 \leqslant 2(\tfrac{1}{4}\epsilon + h^2) + 2(\tfrac{1}{4}\epsilon + h^2) - 4h^2 = \epsilon$$

Thus y_n is a Cauchy sequence and there exists $y \in \mathcal{H}$ such
that $\|y_n - y\| \to 0$. Since M is closed, $y \in M$, and $\|y - x\|$
$= h$ by continuity of the inner product. It remains to show
that $x - y \perp M$. Suppose not. Then there exists $w \in M$ such
that

$$(x - y, w) = r > 0.$$

Now for any $c, y + cw \in M$ so that by (2.19)

$$\|x - y - cw\| \geqslant h = \|x - y\|.$$

Thus

$$0 \leqslant \|x - y - cw\|^2 - \|x - y\|^2$$

$$= \|x - y\|^2 - 2c(x - y, w) + c^2 \|w\|^2 - \|x - y\|^2$$

This shows that

$$2(x - y, w) = 2r < c \|w\|^2$$

But c was arbitrary, so we obtain a contradiction by taking, say, $c = r/\|w\|^2$. Thus $z = x - y$ must be in $M^\perp$. This completes the proof.

We can now introduce the idea of a 'coordinate system' in $\mathcal{H}$. An *orthonormal* set in $\mathcal{H}$ is a set $\Xi = \{\xi_1, \xi_2, \ldots\}$ such that $\|\xi_i\| = 1$ for all i and $\xi_i \perp \xi_j$ for $i \neq j$. Ξ is a *complete* o.n. set, or *orthonormal basis* if $\mathcal{H} = \mathcal{L}(\Xi)$. This is equivalent to saying that if $x \in \mathcal{H}$ then there exists a sequence of real numbers α_i such that $\|x - x_n\| \to 0$ as $n \to \infty$, where

$$x_n = \sum_{i=1}^{n} \alpha_i \xi_i$$

Now by the Schwarz inequality

$$|(\xi_j, x - x_n)| \leqslant \|\xi_j\| \, \|x - x_n\| = \|x - x_n\|$$

But for $n > j$, $(\xi_j, x_n) = \alpha_j$ so that

$$|\alpha_j - (\xi_j, x)| \leqslant \|x - x_n\|$$

which implies $\alpha_j = (\xi_j, x)$ since $\|x - x_n\| \to 0$ as $n \to \infty$. Thus each $x \in \mathcal{H}$ has the representation

$$x = \sum_{i=1}^{\infty} (x, \xi_i) \xi_i$$

This expansion is a generalization of the idea of Fourier series, as will be seen below.

2.2.5. Proposition. *If Ξ is a complete o.n. set in $\mathcal{H}$ and $x, y \in \mathcal{H}$ then*

$$(x, y) = \sum_{i=1}^{\infty} (x, \xi_i)(y, \xi_i). \tag{2.22}$$

In particular

$$\|x\|^2 = \sum_{i} (x, \xi_i)^2 \qquad \textit{(Parseval equality)}.$$

Proof. Suppose $x_n \to x$ in $\mathcal{H}$. Then from Proposition 2.2.2, $\|x_n\| \to \|x\|$. Taking x_n as in (2.21) we have

$$\| x_n \|^2 = \left(\sum_1^n \alpha_i \xi_i, \sum_1^n \alpha_j \xi_j \right) = \sum_1^n \alpha_i^2$$

Thus $\sum_1^n \alpha_i^2 = \sum_1^n (x, \xi_i)^2 \to \| x \|^2$ as $n \to \infty$. The general formula (2.22) now follows from the equality

$$(x, y) = \tfrac{1}{4}(\| x + y \|^2 - \| x - y \|^2).$$

If X is any metric space — that is, a set with a distance function d satisfying properties (2.17) — then a subset $A \subset X$ is *dense* if any element $x \in X$ can be approximated arbitrarily closely by elements of A, i.e. if for any $\epsilon > 0$ there exists $a \in A$ such that $d(x, a) < \epsilon$. X is said to be *separable* if it has a countable dense subset $A = \{a_1, a_2 \ldots \}$.

If B is any subset of a Hilbert space $\mathcal{H}$ then any $x \in \mathcal{H}$ has the decomposition $x = y + z$ where $y \in \mathcal{L}(B)$ and $z \perp \mathcal{L}(B)$, and for any $b \in B$, $d(b, x) \geqslant \| z \|$. It follows that

$$B' = \left\{ x \in \mathcal{H} : x = \sum_{i=1}^n \alpha_i b_i \text{ for some } n, \alpha \in R^n, b_1 \ldots b_n \in B \right\}$$

is dense if and only if $x \perp b$ for all $b \in B$ implies $x = 0$.

Any countable set $\{a_i\}$ in $\mathcal{H}$ can be orthogonalized; using the projection theorem we can apply the Gram-Schmidt procedure as in Proposition 1.1.2 to produce a sequence $\{x_1, x_2 \ldots \}$ such that $\|x_i\| = 1$, $x_i \perp x_j$ for $i \neq j$ and $\mathcal{L}\{x_1, x_2, \ldots \} = \mathcal{L}\{a_1, a_2, \ldots \}$. This means that $\mathcal{H}$ is separable if and only if $\mathcal{H}$ has an o.n. basis. Indeed, if $\{\xi_i\}$ is an o.n. basis then

$$B = \left\{ x \in \mathcal{H} : x = \sum_{i=1}^n r_i \xi_i \text{ for some } n \text{ and rational } r_1 \ldots r_n \right\}$$

is a countable dense subset. Conversely if $\{a_i\}$ is a countable dense subset then for any $x \in \mathcal{H}$ we have $x = \lim_n a_{k_n}$, where a_{k_n} is chosen for example so that $d(x, a_{k_n}) < 1/n$. If $\{x_i\}$ is the orthogonalized sequence corresponding to $\{a_i\}$ then each a_{k_n} is a finite sum

$$a_{k_n} = \sum_i c_{ni} x_i.$$

Since $a_{k_n} \to x$, $(a_{k_n}, x_i) \to (x, x_i)$ (Proposition 2.2.2), i.e.

$c_{ni} \to (x, x_i)$ and it follows that

$$x = \lim_n \sum_{i=1}^{n} (x, x_i) x_i$$

so that $\{x_i\}$ is an o.n. basis.

The fundamental example of a Hilbert space possessing an o.n. basis is $L_2[0, 1]$.

2.2.6. Proposition. $L_2[0, 1]$ *is separable.*

Proof*. First, the set of simple functions, i.e. those of the form

$$f(t) = \sum_i c_i I_{F_i}(t)$$

where the c_i are constants and F_i are measurable sets, is dense in $L_2[0, 1]$. Indeed, for any $z \in L_2$, $F_n = \{t : z(t) > 1/n\}$ must have positive measure for some n unless $z(t) \leqslant 0$ a.s. Denoting the Lebesgue measure of F_n by $\lambda(F_n)$, we then have

$$\int_0^1 z(t) I_{F_n}(t) \, dt > \frac{1}{n} \int_0^1 I_{F_n} dt = \frac{1}{n} \lambda(F_n) > 0.$$

Applying the same argument to $-z$ we see that if $(z, I_F) = 0$ for all measurable sets then $z = 0$ a.s. According to the criterion mentioned above this means that simple functions are dense.

The next step is to show that $z = 0$ if $(z, I_G) = 0$ for all *intervals* $G = (a, b) \subset [0, 1]$. This is due to the way in which Lebesgue measure is constructed and in particular to the fundamental result [4, Theorem 4.5] which says that for any measurable set F and $\epsilon > 0$ there is an open set $G \supset F$ such that $\lambda(G - F) < \epsilon$. Now suppose $(z, I_{(a,b)}) = 0$ for all intervals (a, b); then $(z, I_G) = 0$ for all open sets G since these are just countable unions of intervals. If F is any measurable set and G an ϵ-approximating open set as above then by using the Schwarz inequality (Proposition 2.2.1(a)) we see that

$$|(z, I_F)| = |(z, I_G - I_F)|$$
$$\leqslant \|z\| \, \|I_{G-F}\|$$

$$= \|z\| \sqrt{\lambda(G-F)}$$
$$\leqslant \|z\| \sqrt{\epsilon}.$$

Thus $(z, I_F) = 0$. Finally, since $I_{(a,b)} = \lim_{n} I_{(a_n, b_n)}$ where $\{a_n\}, \{b_n\}$ are sequences of dyadic rational numbers (numbers of the form $k/2^m$ for integral k, m) converging to a, b respectively, it suffices to check that $(z, I_{(a,b)}) = 0$ for all dyadic rational a, b in order to ensure that $z = 0$. But this means that the countable set

$$\left\{ f \in L_2[0, 1] : f = \sum_{i=1}^{n} r_i I_{(a_i, b_i)} \right.$$

$$\left. \text{for some } n, [r_i, a_i, b_i] \text{ dyadic rational} \right\}$$

is dense in $L_2[0, 1]$. This completes the proof.

Of course, o.n. bases are not unique and in $L_2[0, 1]$ there are several convenient ones, perhaps the best known being the trigonometric functions $\{\xi_0, \xi_1, \eta_1, \xi_2, \eta_2 \ldots \}$ where

$$\xi_0(t) = 1, \xi_n(t) = \sqrt{2} \sin 2n\pi t, \eta_n(t) = \sqrt{2} \cos 2n\pi t.$$

It is easily checked that these are orthonormal; an argument to show their completeness is outlined in [4, Exercise 8.3.2]. This shows that any $f \in L_2[0, 1]$ can be expanded in a Fourier series. Another basis, for which the completeness argument is particularly simple, consists of the *Haar functions*

$$\{f_0, \{f_{k/2^n}, k = 1, 3, \ldots 2^n - 1\}, n = 1, 2 \ldots \}$$

where

$$f_0(t) \equiv 1$$

and

$$f_{k/2^n}(t) = \begin{cases} + 2^{(n-1)/2} & (k-1)2^{-n} \leqslant t < k2^{-n} \\ - 2^{(n-1)/2} & k2^{-n} \leqslant t < (k+1)2^{-n} \\ 0 & \text{otherwise} \end{cases}$$

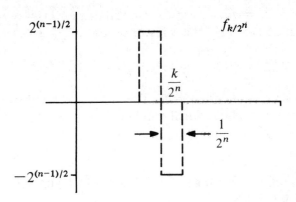

Fig. 2.2

These are handle-shaped objects as shown in Fig. 2.2. It is easy to see that they are mutually orthogonal, i.e.

$$\int_0^1 f_\alpha(t) f_\beta(t) \, dt = 0 \qquad \text{for } \alpha \neq \beta$$

and the value $2^{(n-1)/2}$ is chosen simply to ensure that $\| f_{k/2^n} \| = 1$.

2.2.7. Proposition. *The Haar functions are complete in* $L_2[0, 1]$.

Proof. Take $z \in L_2[0, 1]$ and suppose $z \perp f_0$ and $z \perp f_{k/2^n}$ for all n and k odd. Fix n and define

$$J_i = \int_{i/2^n}^{(i+1)/2^n} z(t) \, dt \qquad i = 0, 1 \ldots 2^n - 1$$

Now $z \perp f_{1/2^n}$, i.e.

$$0 = (z, f_{1/2^n}) = \int_0^1 z(t) f_{1/2^n}(t) \, dt = 2^{(n-1)/2}(J_0 - J_1)$$

Thus $J_0 = J_1$. Similarly $z \perp f_{3/2^n}$ implies $J_2 = J_3$, and so on. But we also have $z \perp f_{k/2^{n-1}}$. For $k = 1$ this says

$$0 = (z, f_{1/2^{n-1}}) = 2^{(n-2)/2}(J_0 + J_1 - J_2 - J_3)$$

and hence $J_0 = J_1 = J_2 = J_3$. Continuing in this way we find that all the J_i are the same. But now $z \perp f_0$ so that

$$0 = \int_0^1 z(t)f_0(t)dt = \int_0^1 z(t)dt = 2^n J_0$$

and hence $J_0 = J_1 = \ldots = 0$. The above argument shows that for any dyadic rational numbers a, b,

$$\int_b^a z(t)dt = 0$$

and this implies $z = 0$ a.s. according to the argument of Proposition 2.2.6.

2.3. Spaces of Square-Integrable Random Variables

Let $(\Omega, \mathscr{B}, P)$ be a probability space. A random variable X on $(\Omega, \mathscr{B}, P)$ is square-integrable if $EX^2 < \infty$, and we recall from Section 2.1 that this implies $E|X| < \infty$ also. The set $\tilde{\mathcal{H}}$ of all such r.v.'s will, with slight modifications, be a Hilbert space. The algebraic operations are defined in the obvious way and the inner product is given by

$$(X, Y) = EXY.$$

It follows that the distance between X and Y is

$$d(X, Y) = \|X - Y\| = \{E(X - Y)^2\}^{1/2}.$$

A sequence X_n converges to X if $d(X_n, X) \to 0$, i.e. if $E(X_n - X)^2 \to 0$. This is called *convergence in quadratic mean* (q.m.). Here we run into a minor difficulty: if the Hilbert space axioms (2.15) are to be satisfied, $d(X, Y) = 0$ should imply that $X = Y$. However, according to Proposition 2.1.3(c), $E(X - Y)^2 = 0$ only implies $X = Y$ a.s., so that it is possible for X and Y to differ on a set D as long as $PD = 0$. This is a technicality, and we get round it by agreeing to regard such r.v.'s as identical. Formally, we consider, not $\tilde{\mathcal{H}}$ but the space $\mathcal{H}$ of *equivalence classes* of r.v.'s in $\tilde{\mathcal{H}}$, two r.v.'s X and Y being equivalent if $P(X = Y) = 1$. Note that, by Proposition 2.1.3(a), if X, X' and Y, Y' are two pairs of equivalent r.v.'s then $(X, Y) = (X', Y')$ so that the inner product is unambiguously defined for elements of $\mathcal{H}$. For the remainder of this book the statement '$X = Y$' where X and Y are r.v.'s always means that they are equivalent. The space $\mathcal{H}$ is often denoted by $L_2(\Omega, \mathcal{B}, P)$ or just L_2 if the underlying probability space is not in doubt. In general L_p is the set of r.v.'s with $E|X|^p < \infty$.

2.3.1. Theorem. $\mathcal{H}$ *is a Hilbert space.*

That $\mathcal{H}$ with the given inner product satisfies the algebraic requirements is more or less immediate, but to prove completeness is more difficult and the reader is referred to Theorem 7.3 of [4] for this.

Two types of convergence for r.v.'s have now been introduced, namely a.s. convergence and q.m. convergence. Neither of these implies the other: it is possible to produce sequences $\{X_n\}$ of r.v.'s which converge a.s. but not in q.m. and conversely. For this purpose it is useful to consider the 'unit interval probability space' where $\Omega = [0, 1]$, $\mathcal{B}$ is the Borel sets of $[0, 1]$ and P is Lebesgue measure, that is the measure that attaches to each interval (a, b) its length $|b - a|$. In this case $\mathcal{H}$ is usually denoted simply as $L_2[0, 1]$ and is the set of square integrable functions on $[0, 1]$ considered in Section 2.2. Then the Haar functions of Proposition 2.2.7 are r.v.'s. The values taken by $f_{k/2^n}$ were chosen so that $\|f_{k/2^n}\| = 1$; if we define $\tilde{f}_{k/2^n} = 2^{-(n-1)/2} f_{k/2^n}$ then $\|\tilde{f}_{k/2^n}\| = 2^{-(n-1)}$ so that $\tilde{f}_{k/2^n} \to 0$ in q.m. as $k, n \to \infty$. However $\tilde{f}_{k/2^n}$ does not converge a.s. since for any $t \in [0, 1]$ and any n_0 there is an $n > n_0$ and k such that $\tilde{f}_{k/2^n}(t) = 1$. The converse situation obtains for the functions $\phi_n = f_{1/2^n}$. Here $\|\phi_n\| = 1$ so that $\phi_n \not\to 0$ in q.m., but $\phi_n \to 0$ a.s. since, for any $t \in [0, 1]$, $\phi_n(t) = 0$ for all $n > 1 - \log_2 t$. This illustrates the following general fact [4, Theorems 7.2, 7.3].

2.3.2. Proposition. *Suppose $\{X_n\}$ is a sequence in $\mathcal{H}$ and $X_n \to X$ in q.m. (i.e. $\|X_n - X\| \to 0$). Then there exists a*

subsequence $\{X_{n_k}, k = 1, 2 \ldots\}$ *such that* $X_{n_k} \to X$ *a.s. as*
$k \to \infty$.

Let

$$\mathcal{H}_0 = \{X \in L_2(\Omega, \mathcal{B}, P) \mid EX = 0\}.$$

Then clearly $\mathcal{H}_0$ is a subspace of $\mathcal{H} = L_2$ and hence a Hilbert
space in its own right. Note that the orthogonal complement
$\mathcal{H}_0^\perp$ of $\mathcal{H}_0$ in $\mathcal{H}$ is one-dimensional and is spanned by the
function $1(\omega) \equiv 1$. Indeed clearly $1 \perp \mathcal{H}_0$ since $E1X = EX$
$= 0$ for $X \in \mathcal{H}_0$, and every $X \in \mathcal{H}$ has the decomposition

$$X = (X - EX) + EX \cdot 1.$$

2.3.3. Proposition. *Suppose* $X_n \to X$ *in* $\mathcal{H}$, *i.e.* $E(X_n - X)^2$
$\to 0$ *as* $n \to \infty$. *Then* $EX_n \to EX$ *and* $\mathrm{var}(X_n) \to \mathrm{var}(X)$.

Proof. If $X_n \to X$ then $(X_n - EX_n) \to (X - EX)$ and $EX_n \to$
EX since these are just the projections onto the orthogonal
subspaces $\mathcal{H}_0$ and $\mathcal{H}_0^\perp$. Now $\mathrm{var}(X_n) = \| X_n - EX_n \|^2$ and
this converges to $\| X - EX \|^2$ by Proposition 2.2.2.

Now suppose we have a stochastic process $\{X_t, t \in R^+\}$ on
some probability space $(\Omega, \mathcal{B}, P)$ such that for each t,

$$EX_t = 0, EX_t^2 < \infty. \tag{2.23}$$

A process satisfying (2.23) is sometimes called a *second-order*
process. Each X_t is an element of the space $\mathcal{H}_0$, so that the
process $\{X_t\}$ can be regarded as a *curve* (a one-parameter
family of points) in a Hilbert space. This curve is continuous
if $\{X_t\}$ is q.m. continuous, i.e. if $E(X_t - X_s)^2 \to 0$ as $s \to t$.
There is a family of subspaces $\mathcal{H}_t^X \subset \mathcal{H}_0$ corresponding to
$\{X_t\}$, defined by

$$\mathcal{H}_t^X = \mathcal{L}\{X_s, 0 \leqslant s \leqslant t\}.$$

This consists of all linear combinations $\Sigma\alpha_i X_{t_i}$ where $t_i \leqslant t$,
and q.m. limits of such combinations. Notice that $\mathcal{H}_t^X$ is
increasing:

$$\mathcal{H}_t^X \subset \mathcal{H}_{t'}^X \subset \mathcal{H}^X = \mathcal{L}\{X_s, s \geqslant 0\} \quad \text{for} \quad t < t'.$$

Recall that since $\mathcal{H}_t^X$ is a subspace, any $Y \in \mathcal{H}_0$ has the unique decomposition

$$Y = Y_1 + Y_2$$

where $Y_1 \in \mathcal{H}_t^X$ and $Y_2 \perp \mathcal{H}_t^X$. We denote by $\mathcal{P}_t^X$ (or simply by $\mathcal{P}_t$) the projection operator which takes each element into its projection onto $\mathcal{H}_t^X$, i.e.

$$\mathcal{P}_t^X Y = Y_1.$$

It is a projection since clearly $(\mathcal{P}_t)^2 Y = \mathcal{P}_t(\mathcal{P}_t Y) = \mathcal{P}_t Y$. It also has the property that for $s < t$,

$$\mathcal{P}_s Y = \mathcal{P}_s \mathcal{P}_t Y \tag{2.24}$$

This is because $Y = \mathcal{P}_t Y + (Y - \mathcal{P}_t Y)$ and $(Y - \mathcal{P}_t Y) \perp \mathcal{H}_t^X$; but $\mathcal{H}_s^X \subset \mathcal{H}_t^X$ so *a fortiori* $(Y - \mathcal{P}_t Y) \perp \mathcal{H}_s^X$. (2.24) shows that one can project 'in stages' instead of 'all at once', a result analogous to the idea of 'iterated conditional expectations'.

From Theorem 2.2.4 we know that $\mathcal{P}_t Y$ has the property that

$$\| Y - \mathcal{P}_t Y \| = \min_{Z \in \mathcal{H}_t^X} \| Y - Z \|.$$

Since $\| Y - Z \|^2 = E(Y - Z)^2$, this says that $\mathcal{P}_t Y$ is the linear least-squares estimator of Y given $\{X_s, 0 \leqslant s \leqslant t\}$. Thus we have, in principle, solved the linear estimation problem for continuous time stochastic processes: as in the finite-dimensional case, the best estimator of a r.v. Y is simply its projection onto the subspace spanned by the observations.[†] Note that for $t > s$

$$(Y - \mathcal{P}_t Y) \perp (\mathcal{P}_t Y - \mathcal{P}_s Y)$$

since $(\mathcal{P}_t Y - \mathcal{P}_s Y) \in \mathcal{H}_t^X$. Now we can write

$$Y - \mathcal{P}_s Y = (Y - \mathcal{P}_t Y) + (\mathcal{P}_t Y - \mathcal{P}_s Y)$$

and hence

$$\| Y - \mathcal{P}_s Y \|^2 = \| Y - \mathcal{P}_t Y \|^2 + \| \mathcal{P}_t Y - \mathcal{P}_s Y \|^2$$
$$\geqslant \| Y - \mathcal{P}_t Y \|^2.$$

[†] In the non-zero mean case, the best affine estimator of Y is its projection onto the subspace of $\mathcal{H}$ spanned by the observations together with the function 1.

This expresses the obvious fact that the estimation error $\| Y - \mathscr{P}_t Y \|^2$ decreases with increasing t. If, for some t_1, $Y \in \mathcal{H}_{t_1}^X$, then of course $Y = \mathscr{P}_{t_1} Y$. A case in point is where $Y = X_{t_1}$; then, for $t < t_1$, $\mathscr{P}_t X_{t_1}$ is the best linear *prediction* of X_{t_1} given $\{X_s, s \leq t\}$. Does $\mathscr{P}_t X_{t_1}$ converge to X_{t_1} as $t \uparrow t_1$? To answer such questions we have to consider the continuity properties of the process $\{X_t\}$.

Quadratic-mean continuity

The covariance function $r(t, s)$ of a second-order process $\{X_t\}$ was introduced in Section 2.1 and is given by

$$r(t, s) = EX_t X_s = (X_t, X_s).$$

If Y and Z are finite linear combinations of the form

$$Y = \Sigma \alpha_i X_{t_i}, \quad Z = \Sigma \beta_i X_{t_i} \tag{2.25}$$

then evidently

$$(Y, Z) = \sum_{i,j} \alpha_i \beta_j r(t_i, t_j).$$

Now consider the subspace $\mathcal{H}^X$ spanned by the process $\{X_t\}$. Being a subspace, this is a Hilbert space in its own right, and if U and V are any two elements of it then by definition we can write

$$U = \lim Y_n, \quad V = \lim_n Z_n,$$

where $\{Y_n\}, \{Z_n\}$ are sequences of linear combinations of the form (2.16), so that, using Proposition 2.2.2,

$$(U, V) = \lim_n (Y_n, Z_n).$$

Thus (U, V) is determined by the covariance function of the process $\{X_t\}$. This means that the entire structure of the space $\mathcal{H}^X$ is determined by the covariance function $r(t, s)$, and other information about the distributions of the process is irrelevant; consequently the answer to any question relating to the Hilbert space properties of the process must be related to properties of $r(t, s)$. In the case of q.m. continuity this is as follows.

2.3.4. Proposition. *(a)* $\{X_t\}$ *is q.m. continuous at t if and only if* $r(\ .\ ,\ .\ )$ *is continuous at the diagonal point* (t, t).

(b) If $\{X_t\}$ *is q.m. continous for all t then* $r(\ .\ ,\ .\ )$ *is continuous at every point* $(s, t) \in (R^+)^2$.

Proof. If r is continuous at (t, t) then

$$E(X_{t+h} - X_t)^2 = EX_{t+h}^2 + EX_t^2 - 2EX_t X_{t+h}$$

$$= (r(t + h, t + h) - r(t, t)) - 2(r(t, t + h) - r(t, t))$$

$$\to 0 \quad \text{as} \quad h \to 0.$$

Conversely, for any t, t' we can write

$$r(t + h, t' + h') - r(t, t') = EX_{t+h}X_{t'+h'} - EX_t X_{t'}$$

$$= E(X_{t+h}(X_{t'+h'} - X_{t'})) + E(X_{t'}(X_{t+h} - X_t))$$

and an application of the Schwarz inequality shows that the right hand side converges to 0 as $h, h' \to 0$, as long as $\{X_t\}$ is q.m. continuous at t and at t'. The remaining assertions follow.

Remark

Suppose $r(t, t)$ is continuous at *all* $t \in R^+$. Then $\{X_t\}$ is q.m. continuous at all t and hence $r(s, t)$ is actually continuous at all (s, t), not just along the diagonal. Since every non-negative definite function is a covariance function, this is a property of non-negative definitive functions.

We can now answer the question posed at the end of the previous paragraph. Suppose $\{X_t\}$ is a q.m. continuous process and $Y \in \mathcal{H}_t^X$, for some t. Given any $\epsilon > 0$ we can choose $(s_1 \ldots s_n, \alpha_1 \ldots \alpha_n)$ such that

$$\| Y - \sum_{i=1}^n \alpha_i X_{s_i} \| < \frac{1}{2} \epsilon.$$

Now if $s_n = t$ we can choose $s_{n-1} \leqslant t' < t$ such that

$$\| X_{t'} - X_t \| < \epsilon/2|\alpha_n|$$

But then

$$\left\| Y - \left(\sum_{i=1}^{n-1} \alpha_i X_{s_i} + \alpha_n X_{t'} \right) \right\| \leqslant \left\| Y - \sum_{i=1}^{n} \alpha_i X_{s_i} \right\| + \left\| \alpha_n X_t - \alpha_n X_{t'} \right\|$$
$$< \epsilon.$$

Denote the quantity in parentheses on the left by Z. Then $Z \in \mathcal{H}_{t'}^X$, $t' < t$, and $\| Y - Z \| < \epsilon$. But by the minimum-norm property $\| Y - \mathcal{P}_{t'} Y \| \leqslant \| Y - Z \|$. Thus $\mathcal{P}_{t'} Y \to Y$ as $t' \uparrow t$.

Typically a process fails to be q.m. continuous when it has discontinuities at fixed times. For example, consider the process $\{X_t\}$ defined by

$$X_t(\omega) = \begin{cases} 0, & t < 1 \\ Z(\omega), & t \geqslant 1 \end{cases}$$

where Z is, say, a $N(0, 1)$ r.v. Then $\{X_t\}$ is obviously not q.m. continuous. In fact $H_t^X = \{0\}$ for $t < 1$ and $H_t^X = \mathcal{L}(Z) = \{\alpha Z : \alpha \in R\}$ for $t \geqslant 1$. On the other hand, q.m. continuity does not mean that the sample functions of the process are continuous. For example, we shall show in Section 3.2 that for the Poisson process $\{N_t\}$ of Definition 2.1.2, $E(N_t - N_s)^2 = (t - s)$, so that $\{N_t\}$ is q.m. continuous. The point is that the discontinuities of its sample paths occur at random times, and the probability that one of them occurs in the interval (s, t) is small for $|t - s|$ small.

Finally, notice that if $\{X_t\}$ is q.m. continuous then the Hilbert space $\mathcal{H}^X$ is separable, because if Q denotes the rational numbers of R^+ then $\{X_r, r \in Q\}$ is a countable set and $\mathcal{H}^X = \mathcal{L}\{X_r, r \in Q\}$ since for any $t \in R^+$, $X_t = \lim_{\substack{r \to t \\ r \in Q}} X_r$.

Quadratic mean integrals

Suppose $\{X_t\}$ is a q.m. continuous process. We shall often want to calculate integrals of the form

$$Y(\omega) = \int_a^b g(t) X_t(\omega) \mathrm{d}t \qquad (2.26)$$

i.e. Lebesgue integrals of the t-function $g(\cdot)X(\cdot, \omega)$ for each fixed ω, thus creating a new r.v. Y. A difficulty arises here in that the process $\{X_t\}$ has been defined simply as a collection of r.v.'s defined for each t, and there is nothing in the definition which assures us that when we fix ω and allow t to vary, we obtain a measurable function of t, as required by the definition of the Lebesgue integral (see Section 2.1). However there is a certain amount of leeway in that we do not mind replacing X_t by another r.v. $\tilde{X}_t$ so long as $P[X_t = \tilde{X}_t] = 1$, so the problem is to take advantage of this intrinsic lack of uniqueness to piece together a measurable function $\tilde{X}(t, \omega)$ such that $P[\tilde{X}_t(\omega) = X_t(\omega)$ for all $t] = 1$. Then $\{\tilde{X}_t\}$ is called a *measurable version* of $\{X_t\}$. It is beyond our scope to go into such questions here: we merely quote the following result, which is a special case of [3], Theorem 2.6.

2.3.5. Proposition. *There is a measurable version of every q.m. continuous process.*

In the sequel it will be tacitly assumed that the measurable version is the one referred to. Then integrals such as (2.26) make sense. Now (2.26) represents a linear operation in X, and the important thing to notice is that, for suitable functions g, $Y \in \mathcal{H}^X$. The precise result is as follows.

2.3.6. Proposition. *Let $g:[a, b] \to R$ be a measurable function such that*

$$\int_a^b g^2(s)ds < \infty \qquad (2.27)$$

and $\{X_t\}$ a q.m. continuous process. Then the r.v. Y defined by (2.26) is in $\mathcal{H}^X$ and

$$\| Y \|^2 = \int_a^b \int_a^b g(t)g(s)r(t, s)ds\,dt \qquad (2.28)$$

where $r(t, s) = \text{cov}(X_t, X_s)$.

Proof*. Since $\{X_t\}$ is q.m. continuous, $r(\,.\,,\,.\,)$ is continuous on $[a, b]$ and hence bounded by some constant K. Thus

$$\int_a^b \int_a^b |g(t)g(s)r(t,s)| \, ds dt \leqslant K \int_a^b \int_a^b |g(t)g(s)| \, ds dt$$

$$= K \left(\int_a^b |g(t)| \, dt \right)^2 < \infty$$

so that the integral on the right in (2.28) is well-defined.

First, consider the Riemann sums corresponding to $\int_a^b X_s ds$. For $n = 1, 2 \ldots$ let $a = t_0^n < t_1^n \ldots < t_{k_n}^n = b$ be a partition of $[a, b]$ such that $\max_i |t_i^n - t_{i-1}^n| \to 0$ as $n \to \infty$, and let

$$Y_n(\omega) = \sum_{i=0}^{k_n - 1} X_{t_i^n}(\omega)(t_{i+1}^n - t_i^n)$$

then

$$EY_n^2 = E \left(\sum_i X_{t_i^n}(\omega)(t_{i+1}^n - t_i^n) \right) \left(\sum_j X_{t_j^n}(\omega)(t_{j+1}^n - t_j^n) \right)$$

$$= \sum_{i,j} r(t_i^n, t_j^n)(t_{i+1}^n - t_i^n)(t_{j+1}^n - t_j^n).$$

Since $r(t, s)$ is continuous we can see that, as $n \to \infty$,

$$EY_n^2 \to \int_a^b \int_a^b r(t, s) \, dt ds.$$

An exactly similar argument shows that $EY_n Y_m$ converges to the same limit as $n, m \to \infty$. Denoting the limit by A we therefore get

$$E(Y_n - Y_m)^2 = EY_n^2 - 2EY_n Y_m + EY_m^2 \to A - 2A + A = 0$$

as $n, m \to \infty$, so that $\{Y_n\}$ is a Cauchy sequence in $\mathcal{H}^X$, with limit, say $\tilde{Y}$, which satisfies (2.28) with $g \equiv 1$. Now Proposition 2.3.2 asserts that every q.m. convergent sequence has an a.s. convergent subsequence, i.e. there is a sequence $n_1 < n_2 < \ldots$ such that $Y_{n_k}(\omega) \to \tilde{Y}(\omega)$ as $k \to \infty$ for all ω except for a set of probability 0. But since the $Y_n(\omega)$ are Riemann sums their limit can only be $\int_a^b X_s(\omega) ds$ (cf. [4],

Theorem 5.9). Thus

$$\tilde{Y}(\omega) = \int_a^b X_s(\omega)\,\mathrm{d}s \qquad \text{a.s.}$$

If g is a step function, i.e.

$$g(t) = \sum_{i=1}^n c_i I_{[t_{i-1}, t_i)}(t)$$

for some constants c_i and $a = t_0 < t_1 .. < t_n = b$, then

$$Y = \int_a^b g(t)X_t\,\mathrm{d}t = \sum_{i=1}^n c_i \int_{t_{i-1}}^{t_i} X_t\,\mathrm{d}t$$

so that $Y \in \mathcal{H}^k$ and satisfies (2.28) by a similar argument to the above. Now recall from the proof of Proposition 2.2.6 that such step functions are dense in $L_2[a, b]$, so that for any g satisfying (2.27) there is a sequence $\{g_n\}$ of step functions such that

$$\int_b^a (g_n(t) - g(t))^2\,\mathrm{d}t \to 0, \quad n \to \infty.$$

Then, as above,

$$E\left[\int_a^b g_n(t)X_t\,\mathrm{d}t - \int_a^b g_m(t)X_t\,\mathrm{d}t \right]^2 = E\left[\int_a^b (g_n(t) - g_m(t))X_t\,\mathrm{d}t \right]^2$$
$$\leqslant K \int_a^b (g_n(t) - g_m(t))^2\,\mathrm{d}t.$$

Thus $\int_a^b g_n(t)X_t\,\mathrm{d}t$ is a Cauchy sequence in $\mathcal{H}^x$ whose limit Y satisfies (2.28) by continuity of the inner product.

Proposition 2.3.6 could have been proved simply by applying the Fubini Theorem [4, Theorem 6.5] to the triple integral $(\mathrm{d}s \times \mathrm{d}t \times \mathrm{d}P)$ involved in calculating $\| Y \|$ in (2.28). However the above argument gives the extra information that $\int_a^b X_t\,\mathrm{d}t$ can be defined as a q.m. limit of Riemann sums, *even though the sample functions of $\{X_t\}$ may not be Riemann integrable.* In this sense 'stochastic' integration turns out, somewhat paradoxically, to be simpler than 'ordinary' integration.

The result of Proposition 2.3.6 is still valid if (2.27) is replaced by the weaker condition

$$\int_a^b |g(s)| \, ds < \infty$$

(use the Fubini Theorem) but we shall not need this generalization.

Hilbert space and normal processes

As remarked above, the structure of the Hilbert subspace $\mathcal{H}^X \subset \mathcal{H}$ corresponding to a second-order process $\{X_t\}$ is entirely determined by the covariance function. Since in the normal case the covariance function also characterizes the distributions of the process, it is clear there must be some relation between these concepts. The basic result is the following.

2.3.7. Proposition. *Suppose $\{X_n\}$ is a sequence of normal r.v.'s converging in q.m. to a r.v. X. Then X is normal.*

Proof. We will show that the sequence of characteristic functions converges. Let $m_n = EX_n$ and $v_n = \text{var}(X_n)$. Then

$$Ee^{iuX_n} = \phi_n(u) = e^{ium_n - \frac{1}{2}v_n u^2}.$$

For $x, y \in R$, let

$$a = \tfrac{1}{2}(x + y), \quad \theta = \tfrac{1}{2}(x - y).$$

Then

$$e^{iux} - e^{iuy} = e^{iua}(e^{iu\theta} - e^{-iu\theta})$$

$$= 2ie^{iua} \sin u\theta$$

and hence

$$|e^{iux} - e^{iuy}| = 2|\sin u\theta| \leqslant 2|u||\theta| = |u||x - y|.$$

This shows that the sequence $\{e^{iuX_n}\}$ is q.m. convergent for each u, since

$$E(e^{iuX_n} - e^{iuX})^2 \leqslant u^2 E(X_n - X)^2.$$

Now from Proposition 2.3.3, $Ee^{iuX_n} \to Ee^{iuX}$, i.e.

$$Ee^{iuX} = \lim_n \phi_n(u).$$

But, again from Proposition 2.3.3, $m_n \to m = EX$ and $v_n \to v = \text{var}(X)$. Thus

$$\phi_n(u) = e^{im_n - \frac{1}{2}v_n u^2} \to e^{imu - \frac{1}{2}vu^2}$$

so that X is normal, as claimed.

The above result has the following implications for the Hilbert space framework: if $\{X_\alpha, \alpha \in I\}$ is any normal family of elements of $L_2(\Omega, \mathscr{B}, P)$ (i.e. $(X_{\alpha_1} \ldots X_{\alpha_n})$ is a normal vector for any choice of $\alpha_1 \ldots \alpha_n \in I$), then every element of $\mathcal{L}\{X_\alpha, \alpha \in I\}$ is also normal, since these elements consist precisely of finite linear combinations and q.m. limits.

Now suppose $\{X_t\}$ is a normal process and Y another normal r.v. Let

$$\hat{Y} = \mathscr{P}_t^X Y$$

i.e. $\hat{Y}$ is the projection of Y onto $\mathcal{H}_t^X$. Then by the above remarks $\hat{Y}$ and $\tilde{Y} = Y - \hat{Y}$ are normal; but $\hat{Y} \perp \tilde{Y}$, i.e. $\hat{Y}$ and $\tilde{Y}$ are uncorrelated. Therefore they are independent. Let us consider the conditional characteristic function $\phi_{Y|X}(u)$ of Y given $\{X_s, s \leqslant t\}$. This is given by

$$\begin{aligned}
\phi_{Y|X}(u) &= E[e^{iuY} | X_s, s \leqslant t] \\
&= E[e^{iu\hat{Y}} e^{iu\tilde{Y}} | X_s, s \leqslant t] \\
&= e^{iu\hat{Y}} E[e^{iu\tilde{Y}}].
\end{aligned}$$

We have used the independence and the fact that $\hat{Y} \in \mathcal{H}_t^X$ in the transition to the last line. Since $\tilde{Y}$ is normal, $Ee^{iu\tilde{Y}} = e^{-\frac{1}{2}\sigma^2 u^2}$ where $\sigma^2 = \text{var}(\tilde{Y})$. Thus

$$\phi_{Y|X}(u) = e^{iu\hat{Y} - \frac{1}{2}\sigma^2 u^2}.$$

But this is the characteristic function of the $N(\hat{Y}, \sigma^2)$ distribution. Thus $\hat{Y}$ is actually the conditional mean of Y given $\{X_s, s \leqslant t\}$, so that, as in the finite-dimensional case, $\hat{Y}$ is the minimum mean-square error estimator of Y given $\{X_s, s \leqslant t\}$, linear or not.

It should be pointed out that the above argument is incomplete in that we have not defined rigorously what is meant by conditional expectations given processes. However, note that the only properties we have used are the following (here X denotes $\{X_s, s \leqslant t\}$)

(1) For any r.v. Z and function g, $E(g(X)Z \mid X) = g(X)E(Z \mid X)$.

(2) If Z is independent of X, $E(Z \mid X) = EZ$.

These must certainly be satisfied for any reasonable definition of conditional expectation; an appropriate formulation will be found in the Appendix.

We have now characterized linear estimation in terms of projection in Hilbert space. The problem that remains is that of actually calculating these projections. In the finite dimensional case we saw that this could be done by using the Gram-Schmidt procedure to construct an orthonormal basis. In principle we could do the same for q.m. continuous processes since they span separable Hilbert subspaces, but this is not generally a viable procedure. To go further we have to introduce more structure into the problem by considering special classes of processes. One class which has both nice mathematical properties, and arises naturally in a variety of physical applications, is the class of *orthogonal increments processes*. These are considered in the next chapter.

2.4. Problems and Complements

1. Since the d.f. of the $N(0, 1)$ distribution cannot be evaluated in closed form, the method suggested by the proof of Proposition 2.1.5 is not a good way of generating $N(0, 1)$ r.v.'s from uniform $[0, 1]$ samples. Such r.v.'s are best generated in pairs. If X_1 and X_2 are independent $N(0, 1)$ then their joint density function is

$$\frac{1}{2\pi} e^{-\frac{1}{2}(x_1^2 + x_2^2)}.$$

It follows that $\theta = \tan^{-1}(X_2/X_1)$ is uniformly distributed on $[0, 2\pi]$ and $R = \sqrt{X_1^2 + X_2^2}$ has density

$$r e^{-1/2 r^2}.$$

Thus if U_1, U_2 are independent, uniform $[0, 1]$, then

$$X_1 = \sqrt{2ln(1/U_1)} \; \cos 2\pi U_2, X_2 = \sqrt{2ln(1/U_1)} \; \sin 2\pi U_2$$

are independent $N(0, 1)$.

(An alternative is to rely on the central limit theorem and take $X = (U_1 + U_2 + \ldots + U_n)/\sqrt{n}$. For most purposes an n of around 12 suffices.)

2. An exponential r.v. X can be generated as $X = ln(1/U)$, but there is another way which avoids function evaluations.

U_1, $U_2 \ldots U_0^*$, $U_1^* \ldots$ are independent uniform $[0, 1]$ r.v.'s.

(a) Fix $t \in (0, 1)$ and define $N(t)$ as follows:

$$\text{if } U_1 \geqslant t, N(t) = 1$$

$$\text{if } U_1 < t, N(t) = \min \{k : U_k > U_{k-1}\}.$$

Then $P[N(t) \text{ is odd}] = e^{-t}$.

(b) Define X by the recursive algorithm

$$X > 0.$$

$$\text{Suppose } X > k. \text{ If } N(U_k^*) \text{ is odd, } X = k + U_k^*.$$

$$\text{If } N(U_k^*) \text{ is even, } X > k + 1.$$

Then $P[X > \alpha] = e^{-\alpha}$.

These two steps can be combined in the single algorithm below. A fresh sample is taken each time the 'get' statement is encountered, and $a : = b$ means 'give a the value b'.

$$k : = 0$$

$$1 : \text{get } U_0$$

$$2 : \text{get } U_1 ; \text{if } U_1 > U_0, X : = U_0 + k; \text{stop.}$$

$$\text{get } U_2 ; \text{if } U_2 > U_1, k : = k + 1; \text{go to 1}$$

$$U_0 : = U_2 ; \text{go to 2.}$$

CHAPTER THREE

Orthogonal increments processes

3.1. General Properties

Throughout this section all r.v.'s are assumed to be defined on some fixed probability space $(\Omega,\ \mathscr{F}, P)$. Further, they will be taken to have zero mean and finite variance, i.e. to belong to the Hilbert space $\mathcal{H}_0 = \{X \in L_2(\Omega,\ \mathscr{B}, P): EX = 0\}$ unless otherwise specified.

3.1.1. Definition. $\{X_t, t \geqslant 0\}$ *is an* orthogonal increments (o.i.) process *if for any non-overlapping intervals* $(s,\ t)$, (s',t')

$$(X_{t'} - X_{s'}) \perp (X_t - X_s). \qquad (3.1)$$

Notice that for any $X \in \mathcal{H}$, the process $\{X_t'\}$ defined by $X_t' = (X_t + X)$ has orthogonal increments if $\{X_t\}$ has. In particular we can take $X = -X_0$, in which case $X_0' = 0$. Thus we will assume henceforth that $X_0 = 0$.

3.1.2. Proposition. *Let* $\{X_t\}$ *be an o.i. process and*

$$\psi(t) = EX_t^2 = \|X_t\|^2.$$

Then ψ *is a non-decreasing function of t and the covariance function $r(t, s) = (X_t, X_s)$ is given by*

$$r(t, s) = \psi(t \wedge s) \qquad (3.2)$$

where $t \wedge s = \min(t, s)$.

Proof. Take $0 < s < t$. Then
$$X_t = (X_t - X_s) + (X_s - X_0).$$
Now $X_0 = 0$ so that, using (3.1),
$$EX_t^2 = E(X_t - X_s)^2 + EX_s^2 \qquad (3.3)$$
and hence
$$\psi(t) - \psi(s) = E(X_t - X_s)^2 \geqslant 0$$
so that ψ is non-decreasing. Also
$$\begin{aligned} r(t, s) &= EX_t X_s \\ &= E(X_s[(X_t - X_s) + X_s]) \\ &= EX_s^2 = \psi(s). \end{aligned}$$
Similarly, $r(t, s) = \psi(t)$ if $0 \leqslant t < s$. Formula (3.2) covers both cases.

A process $\{X_t\}$ has *stationary increments* if the variance of the increment $(X_t - X_s)$ depends only on the distance $|t - s|$, i.e. if for any r, s, t
$$E(X_t - X_s)^2 = E(X_{t+r} - X_{s+r})^2.$$
Thus if $\{X_t\}$ has stationary orthogonal increments then we see from (3.3) that its variance $\psi(t)$ satisfies, for $t > s$
$$\psi(t) = \psi(t - s) + \psi(s)$$
or alternatively
$$\psi(t + r) = \psi(t) + \psi(r), \ \psi(0) = 0$$
for $t, r > 0$. The only continuous solution to this equation is
$$\psi(t) = \sigma^2 t$$
for some constant σ^2 (and of course $\sigma^2 = \psi(1) = \mathrm{var}(X_1)$). We thus have the following result.

3.1.3. Proposition. *A q.m. continuous process* $\{X_t\}$ *has stationary orthogonal increments if and only if its covariance function is*

$$r(t, s) = \sigma^2 t \wedge s$$

for some constant σ^2.

Proof. The 'only if' part follows from the above argument and Proposition 2.3.4. Conversely, if $r(t, s) = \sigma^2 t \wedge s$ then $r(t, t) = \sigma^2 t$ so that $\{X_t\}$ is q.m. continuous by Prop. 2.3.4, and for $s \leqslant t \leqslant s' \leqslant t'$

$$
\begin{aligned}
(X_{t'} - X_{s'}, X_t - X_s) &= (X_{t'}, X_t) - (X_{t'}, X_s) \\
&\quad - (X_{s'}, X_t) + (X_{s'}, X_s) \\
&= \sigma^2 (t' - t' - s' + s') = 0.
\end{aligned}
$$

This completes the proof.

$\{X_t\}$ has *independent increments* if $(X_t - X_s)$, $(X_{t'} - X_{s'})$ are independent r.v.'s for non-overlapping (s, t) and (s', t'). If the variances are finite then this implies that the increments are orthogonal, so that second-order independent increment processes are a sub-class of o.i. processes. Of course to determine whether a given process has independent increments involves knowing more than just the covariance function; but for any o.i. process the equivalent normal process (see Section 2.1) has independent increments.

Here is a general way of constructing o.i. processes. Let $\{\mathcal{H}_t, t \in R^+\}$ be an increasing family of subspaces of $\mathcal{H}$ (for example, $\mathcal{H}_t = \mathcal{L}\{X_s, s \leqslant t\}$ for some process $\{X_t\}$) and let Z be any element of $\mathcal{H}$. Now define

$$Z_t = \mathscr{P}_t Z$$

where $\mathscr{P}_t$ is the projection onto $\mathcal{H}_t$; then the process $\{Z_t\}$ has o.i. Indeed, for $t \leqslant r$,

$$\mathscr{P}_t Z_r = \mathscr{P}_t \mathscr{P}_r Z = \mathscr{P}_t Z = Z_t$$

which does not depend on r, so that $\mathscr{P}_t(Z_{t'} - Z_{s'}) = 0$, i.e. $(Z_{t'} - Z_{s'}) \perp \mathcal{H}_t$, for $t', s' \geqslant t$.

$\{Z_t\}$ generates its own family of subspaces

$$\mathcal{H}_t^Z = \mathcal{L}\{Z_s, s \leqslant t\}$$

and by definition $\mathcal{H}_t^Z \subset \mathcal{H}_t$. Suppose it is possible to find N elements $Z^1 \ldots Z^N \in \mathcal{H}$ such that

$$\mathcal{H}_t = \mathcal{H}_t^{Z^1 \ldots Z^N} = \mathcal{L}\{Z_s^1 \ldots Z_s^N, s \leqslant t\}$$

and that N is the minimum number for which such elements can be found. Then the number N is called the *multiplicity* of the family of subspaces. As will been seen below, very explicit information about the structure of the subspaces generated by o.i. processes is available, and therefore the idea of replacing a process $\{X_t\}$ by 'equivalent' o.i. processes $Z^1 \ldots Z^N$ which generate the same subspaces is attractive where this can be done. Indeed the theory of Kalman filtering, developed in Chapter 4, is based on the fact that in that case the m-vector observed process has multiplicity m.

In the next two sections we introduce some particular classes of orthogonal increments processes before returning to the general development in Section 3.4. In particular we describe the two classic stationary independent increments processes: the Poisson process and Brownian motion. It is worth remarking that these, which look at first sight like very special cases, are in a sense exhaustive; Brownian motion is the *only* such process with continuous sample functions, and it turns out that any stationary independent increment process must be some kind of combination of Brownian motion and Poisson processes. The point is that to insist that the increments be stationary and independent is a very restrictive condition (we have already seen, for example, that essentially only one covariance function is possible) which can only be satisfied in a limited number of ways.

3.2. Counting processes

$\{N_t, t \geqslant 0\}$ is a *counting process* if there is an increasing sequence of r.v.'s $0 < T_1 < T_2 \ldots$ such that

$$N_t = \sum_i I_{(t \geqslant T_i)}.$$

Thus N_t takes values in Z^+ and $N_t = n$ if just n of the T_i's are less than or equal to t, i.e. have 'happened' at times up to and including t. We have already seen an example in the Poisson process of Definition 2.1.1. Counting processes arise in a variety of applications: in statistics or operations research as customers arriving at a queue, demands for telephone service or traffic flow past a check point; in physics as counters of radioactive emissions or photoelectrons; and in many other areas.

Probabilistically, we have to specify the distributions of the r.v.'s $\{T_i\}$, or equivalently of the interarrival times $\{S_i\}$ where

$$S_1 = T_1$$
$$S_k = T_k - T_{k-1} \qquad k \geqslant 2.$$

The simplest case, and the only one we consider here, is where $\{S_i\}$ is an i.i.d. sequence; then the whole probabilistic structure is determined by a single distribution function. If this is exponential we get the Poisson process, which we investigate next.

The Poisson Process

According to Definition 2.1.1, this is a counting process whose interarrival times $\{S_i\}$ are i.i.d. sequence of r.v.'s such that

$$P(S_i > t) = e^{-\lambda t}$$

for some constant λ. The crucial fact about the exponential distribution is its 'forgetting' property:

$$P(S_i > t + s \mid S_i > t) = \frac{e^{-\lambda(t+s)}}{e^{-\lambda t}} = e^{-\lambda s} \qquad (3.4)$$

Thus the conditional distribution of S_i, given that $S_i > t$, is the same as the unconditional distribution of $t + S$ where S is another exactly similar exponential r.v.

3.2.1. Proposition. *(a) The r.v. N_t has Poisson distribution with parameter λt.*

(b) The process $\{N_t\}$ has independent increments.

Proof. Let
$$p_n(t) = P(N_t = n).$$

We can calculate directly p_0 and p_1. For p_0, we have $N_t = 0$ if and only if $S_1 > t$, so that
$$p_0(t) = P(S_1 > t) = e^{-\lambda t}.$$

Now given that $T_1(=S_1) = s$ where $0 \leqslant s \leqslant t, N_t = 1$ if and only if $T_2 > t$, i.e. $S_2 > t - s$; thus
$$P[N_t = 1 \mid S_1 = s] = e^{-\lambda(t-s)}$$

Since the density function of S_1 is $\lambda e^{-\lambda s}$ we get

$$p_1(t) = \int_0^t e^{-\lambda(t-s)} \lambda e^{-\lambda s} ds = \lambda t e^{-\lambda t}. \qquad (3.5)$$

It is possible to continue in this way, but quicker to use the following induction argument. First

$$\begin{aligned}
P[N_t \geqslant 2] &= 1 - p_0 - p_1 \\
&= 1 - e^{-\lambda t} - \lambda t e^{-\lambda t} \\
&= \frac{(\lambda t)^2}{2} + (\lambda t)^3 \left(\frac{1}{3!} - \frac{1}{2!} \right) + \ldots \\
&= o(t) \,^\dagger \qquad\qquad\qquad\qquad (3.6)
\end{aligned}$$

Now consider the interval $[0, t + \delta]$. Writing $\Delta N = N_{t+\delta} - N_t$, we have

$$\begin{aligned}
(N_{t+\delta} = n) &= [(N_t = n) \cap (\Delta N = 0)] \\
&\cup \; [(N_t = n - 1) \cap (\Delta N = 1)] \\
&\cup \; [\bigcup_{k=2}^{n} (N_t = n - k) \cap (\Delta N = k)].
\end{aligned}$$

Now from (3.4)
$$P[\Delta N = 0 \mid T_n, T_n < t] = e^{-\lambda \delta}$$

and hence

$^\dagger o(t)$ is the generic name for any function with the property that $o(t)/t \to 0$ as $t \to 0$.

$$P[(N_t = n) \cap (\Delta N = 0)] = P[\Delta N = 0 \mid N_t = n] P[N_t = n]$$
$$= e^{-\lambda\delta} p_n(t)$$
$$= (1 - \lambda\delta + o(\delta)) p_n(t).$$

Similarly

$$P[(N_t = n - 1) \cap (\Delta N = 1)] = (\lambda\delta + o(\delta)) p_{n-1}(t).$$

And from (3.6),

$$P(\bigcup_{k=2}^{n} (N_t = n - k) \cap (\Delta N = k) = o(\delta).$$

Thus we get

$$p_n(t + \delta) = (1 - \lambda\delta) p_n(t) + \lambda\delta p_{n-1}(t) + o(\delta).$$

Now letting $\delta \downarrow 0$ gives

$$\frac{d}{dt} p_n(t) = \lambda(p_{n-1}(t) - p_n(t))$$

with initial condition $p_n(0) = 0$ for $n \geqslant 1$. This equation is easily solved by the substitution $p_n = e^{-\lambda t} V_n$; then

$$\dot{V}_n = \lambda V_{n-1}.$$

But from (3.5), $V_1 = \lambda t$: hence $V_n(t) = (\lambda t)^n / n!$. Thus we have shown

$$p_n(t) = P(N_t = n) = \frac{(\lambda t)^n}{n!} e^{-\lambda t} \qquad (3.7)$$

so that N_t has the Poisson distribution with parameter λt.

To show that the Poisson process has independent increments, consider non-overlapping intervals (s', t') and (s, t) with $t' \leqslant s$ and suppose we are given that $N_s = n$ and the values $T_1 \ldots T_n$. Then according to the remarks following (3.4), the r.v. $S = T_{n+1} - s$ is exponentially distributed and hence the r.v.'s $S, S_{n+2}, S_{n+3} \ldots$ form a Poisson process starting at s. Hence

$$P(N_t - N_s = k) = E(I_{(N_t - N_s = k)}) = p_k(t - s) \qquad (3.8)$$

On the other hand we also have

$$E(I_{(N_{t'} - N_{s'} = k')} I_{(N_t - N_s = k)} \mid N_t = n, T_1 \ldots T_n)$$
$$= I_{(N_{t'} - N_{s'} = k')} p_k(t - s)$$

and hence

$$P(N_{t'} - N_{s'} = k', N_t - N_s = k) = p_{k'}(t' - s')p_k(t - s).$$

Since this holds for all k, k', we have shown that $N_{t'} - N_{s'}$ and $N_t - N_s$ are independent. This completes the proof.

In view of (3.8), the distribution of any increment $(N_t - N_s)$ is Poisson with parameter $\lambda(t - s)$. Thus in particular $E(N_t - N_s) = \lambda(t - s)$ so that the parameter λ is simply the expected number of events per unit time, i.e. the *rate* of the process. λ also has the following interpretation: from (3.7) and (3.8),

$$P(N_{t+\delta} - N_t = 0) = 1 - \lambda\delta + o(\delta)$$
$$P(N_{t+\delta} - N_t = 1) = \lambda\delta + o(\delta). \tag{3.9}$$

Thus the probability of observing an event in a short interval $(t, t + \delta)$ is just $\lambda\delta$, and there is negligible probability of observing two or more events. The Poisson process is in fact the only independent increments counting process satisfying (3.9), and is commonly introduced that way (see, for example, [2] Section 3.7); but we chose to start from Definition 2.1.1 so as to accord with the stochastic process structure of Section 2.1.

The centralized Poisson process is

$$Y_t = N_t - \lambda t$$

$\{Y_t\}$ is a zero mean process with stationary orthogonal increments. From the properties of the Poisson distribution we know that $\mathrm{var}(N_t) = \mathrm{var}(Y_t) = \lambda t$; so, from Proposition 3.1.3 the covariance functions for Y_t is $r(t, s) = \lambda t \wedge s$.

Renewal processes

In many situations it is reasonable to assume that the inter-arrival times are i.i.d. but it is necessary for them to have some distribution other than exponential. Then we have a *renewal process*, so called because it clearly models the replacement of nominally identical items, such as light bulbs, with a random lifetime. It would certainly be perverse, given

a bulb with a nominal life of 1000 hours, but which has
already been in service for 2000 hours, to expect it to last
another 1000, but this is what is implied by an exponentially-
distributed lifetime, according to (3.4). Renewal processes
are a big subject in their own right (see for example [1]) and
our objective here is limited to showing that a certain orthogonal increment process can be associated with each renewal
process. This then furnishes another class of o.i. processes
to which the results of the following sections can be applied.

Let T be a non-negative r.v. with a density function f, and
let

$$F(t) = \int_t^\infty f(u)\,du$$

so that $1 - F(\cdot)$ is the d.f. of T. Now consider the elementary stochastic process

$$X_t = I_{(t \geqslant T)}$$

Let $dX_t = X_{t+dt} - X_t$; then $dX_t = 0$ unless $T \in (t, t + dt]$
in which case $dX_t = 1$. Thus

$$E[dX_t \mid T \leqslant t] = 0 \tag{3.10}$$

and

$$E[dX_t \mid T > t] = P[T \in (t, t + dt] \mid T > t] \tag{3.11}$$

$$= \frac{f(t)\,dt}{F(t)}.$$

Now define

$$A_t = \int_0^{T \wedge t} \frac{1}{F(s)} f(s)\,ds$$

and

$$Q_t = X_t - A_t.$$

Then from (3.10) and (3.11) we have

$$E[dQ_t \mid T \leqslant t] = E[dQ_t \mid T > t] = 0.$$

This suggests that Q_t has orthogonal increments, since
$E\,dQ = 0$ whatever the 'past' of the process. Let us show this.
The conditional distribution of T given that $T > s$ has density
function

$$f_s(r) = \frac{f(r)}{F(s)} \quad \text{for} \quad r \geqslant s.$$

Thus for $t > s$

$$E[Q_t - Q_s | T > s] = - \int_s^t \frac{1}{F(u)} f(u) du \frac{F(t)}{F(s)}$$

$$+ \int_s^t \left(1 - \int_s^r \frac{1}{F(u)} f(u) du \right) \frac{1}{F(s)} f(r) dr.$$

The two terms correspond to the events $(T > t)$ and $(s \leqslant T \leqslant t)$ respectively. The second term is

$$\frac{1}{F(s)} \int_s^t f(r) dr - \frac{1}{F(s)} \int_s^t \int_s^r \frac{1}{F(u)} f(u) du f(r) dr$$

$$= \frac{1}{F(s)} (F(s) - F(t)) - \frac{1}{F(s)} \int_s^t \frac{1}{F(u)} f(u) \int_u^t f(r) dr du$$

$$= \frac{1}{F(s)} (F(s) - F(t)) - \frac{1}{F(s)} \int_s^t \frac{1}{F(u)} f(u) (F(u) - F(t)) du$$

$$= \frac{F(t)}{F(s)} \int_s^t \frac{1}{F(u)} f(u) du,$$

and consequently

$$E[Q_t - Q_s | T > s] = 0.$$

In particular, taking $s = 0$ gives $EQ_t = 0$. Now by definition $Q_t - Q_s = 0$ if $T \leqslant s$, so that taking $s' \leqslant t' \leqslant s$ we have

$$E[(Q_t - Q_s)(Q_{t'} - Q_{s'})] = E[(Q_t - Q_s) I_{(T > s)} (Q_{t'} - Q_{s'})]$$

$$= E \left(- \int_{s'}^{t'} \frac{1}{F(u)} f(u) du E[Q_t - Q_s | T > s] \right)$$

$$= 0.$$

Thus $\{Q_t\}$ is a zero-mean o.i. process. Notice that since $f(s) = - \dot{F}(s)$ and $F(0) = 1$ we can write A_t as

$$A_t = - \ln (F(t \wedge T))$$

It is sometimes called the *compensating process* for $\{X_t\}$. If we regard T as the first time of a renewal process, then the process restarts at T and we can obtain the corresponding process for the next time by restarting A_t at T as well. Continuing for each successive renewal time we obtain the following result.

3.2.2. Proposition. *Let* $\{X_t, t \geqslant 0\}$ *be a renewal process with renewal and interarrival times* $\{T_i\}$, $\{S_i\}$ *respectively, and let* G *be the d.f. of* S_i. *Suppose* G *is continuous and define the process* $\{A_t, t \geqslant 0\}$ *by*

$$A_t = -\ln [F(S_1)F(S_2) \ldots F(S_{n-1})F(t - T_{n-1})]$$
$$\text{for } t \in [T_{n-1}, T_n) \tag{3.12}$$

where $F(t) = 1 - G(t)$. *Then the process*

$$Q_t = X_t - A_t$$

is a zero mean o.i. process.

Remark

In the preceding argument we assumed G had a density. This is not essential but simplifies the calculations. On the other hand (3.12) is *not* the appropriate formula if G has discontinuities.

3.2.3. Example. Suppose the $\{S_i\}$ are uniform $[0, 1]$ r.v.'s; then for $0 \leqslant t \leqslant 1$

$$F(t) = 1 - t$$

and of course $S_i < 1$ with probability 1. So in this case

$$A_t = \ln ((1 - S_1)(1 - S_2) \ldots (1 - S_{n-1})(1 - t + T_{n-1})).$$

Representative sample functions of A_t and Q_t are shown in Fig. 3.1.

Notice that in general the compensator $\{A_t\}$ is a stochastic process determined by the past of the renewal process $\{X_t\}$. We can now see in what way the Poisson process is a special case. For the Poisson process $F(t) = e^{-\lambda t}$ so that $-\ln F(t) = \lambda t$; applying (3.12) gives

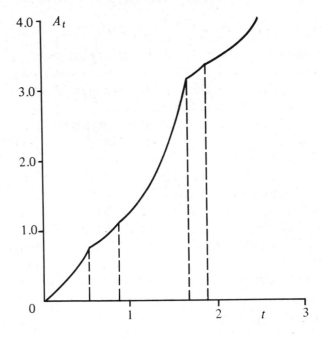

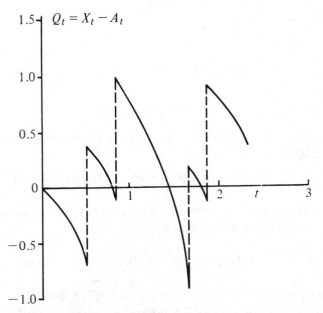

Fig. 3.1

$$A_t = \lambda(S_1 + S_2 + \ldots + S_{n-1} + t - T_{n-1}) = \lambda t.$$

Thus when we piece together A_t in the way suggested by (3.12), the 'joins' are invisible and we obtain a deterministic function $A_t = \lambda t$. It is clear that the exponential is the only continuous d.f. for which this can occur.

We saw in Proposition 3.1.2 that the covariance function — and hence all the Hilbert space properties — of an o.i. process $\{Q_t\}$ is determined by the function $\psi(t) = EQ_t^2$. Let us calculate this for the process $\{Q_t\}$ associated with a renewal process $\{X_t\}$ as in Proposition 3.2.2. The answer is easily stated: it is that $\psi(t) = EX_t$. To show this it is necessary to introduce the *renewal equation*, which also provides a simple formula for calculating EX_t from the d.f. G of the interarrival times.

Suppose, as before, that G has a density function f, and let

$$m(t) = EX_t.$$

Then we can write

$$m(t) = E[E(X_t \mid T_1 = s)]$$

$$= \int_0^\infty E(X_t \mid T_1 = s) f(s) \, ds. \qquad (3.13)$$

If $s > t$ then $X_t = 0$, so that $E(X_t \mid T_1 = s) = 0$. If $s \leqslant t$ then $X_s = 1$ and the process restarts as s; thus

$$E(X_t \mid T_1 = s) = 1 + m(t - s), \quad s \leqslant t.$$

Using this in (3.13) gives the *renewal equation* satisfied by $m(t)$, namely

$$m(t) = G(t) + \int_0^t m(t - s) f(s) \, ds. \qquad (3.14)$$

This equation is easily solved by Laplace transforms (see [1, 9] for full details). If we define

$$m^*(\sigma) = \int_0^\infty e^{-\sigma t} m(t) \, dt, \quad f^*(\sigma) = \int_0^\infty e^{-\sigma t} f(t) \, dt$$

and use the facts that the transform of 1 is $1/\sigma$ and that convolutions transform into products, we obtain from (3.14):

$$m^*(\sigma) = \frac{1}{\sigma} f^*(\sigma) + m^*(\sigma) f^*(\sigma).$$

Thus m^* and f^* are related by

$$m^*(\sigma) = \frac{f^*(\sigma)}{\sigma(1 - f^*(\sigma))}, \quad f^*(\sigma) = \frac{\sigma m^*(\sigma)}{1 + \sigma m^*(\sigma)}$$

(3.15)

Since there is a one-to-one relation between a function and its transform, this solves (3.14), and also shows that f can be calculated from m, i.e. specifying the mean function $m(\cdot)$ is sufficient to determine the interarrival density $f(\cdot)$. For example, if $m(t) = \lambda t$ we find that

$$m^*(\sigma) = \frac{\lambda}{\sigma^2}$$

and

$$f^*(\sigma) = \frac{\lambda}{\sigma + \lambda}$$

so that

$$f(t) = \lambda e^{-\lambda t}.$$

Thus the property that $m(t) = \lambda t$ characterizes the Poisson process. For the Poisson process the compensator $\{A_t\}$ and mean $\{m(t)\}$ coincide, but, as indicated earlier, this is the only renewal process for which this occurs.

The above discussion was restricted, for simplicity, to the case where G has a density. Analogous results apply in general, but Stieltjes integrals and the Laplace–Stieltjes transform must be used; see [9].

We can now calculate the variance for the general renewal process $\{X_t\}$.

3.2.4. Proposition. *For the renewal process described in Proposition 3.2.2,*

$$\psi(t) = m(t)$$

where $\psi(t) = EQ_t^2$ *and* $m(t) = EX_t = EA_t$.

Proof. We are going to show that ψ satisfies the renewal equation (3.14). As above we have

$$\psi(t) = E(X_t - A_t)^2$$

$$= \int_0^\infty E[(X_t - A_t)^2 \mid T_1 = s] f(s) \, ds$$

$$= \int_0^t E[(X_t - A_t)^2 \mid T_1 = s] f(s) \, ds$$

$$+ \int_t^\infty E[(X_t - A_t)^2 \mid T_1 = s] f(s) \, ds. \quad (3.16)$$

Now if $T_1 = s > t$ then $X_t = 0$ and $A_t = -\ln F(t)$. Thus

$$\int_t^\infty E[(X_t - A_t)^2 \mid T_1 = s] f(s) \, ds = F(t)(\ln F(t))^2. \quad (3.17)$$

If $T_1 = s < t$ then

$$Q_s = X_s - A_s = 1 + \ln F(s)$$

and the process restarts at s, so that

$$\int_0^t E[Q_t^2 \mid T_1 = s] f(s) ds = \int_0^t E[Q_{t-s} + 1 + \ln F(s)]^2 f(s) ds$$

$$= \int_0^t \{ EQ_{t-s}^2 + 2E[Q_{t-s}(1 + \ln F(s))]$$

$$+ (1 + \ln F(s))^2 \} f(s) ds$$

$$= \int_0^t \psi(t-s) f(s) ds +$$

$$+ \int_0^t (1 + \ln F(s))^2 f(s) ds. \quad (3.18)$$

(Recall that $EQ_t = 0$.) Now an integration by parts using the fact that $f(t) = -\dfrac{d}{dt} F(t)$ shows that

$$\int_0^t (1 + \ln (F(s))^2 f(s) ds = 1 - F(t) - F(t)(\ln F(t))^2. \quad (3.19)$$

Combining (3.16) – (3.19), we see that ψ satisfies

$$\psi(t) = G(t) + \int_0^t \psi(t-s)f(s)\,ds.$$

But this is the renewal equation (3.14) whose unique solution is $m(t)$. This completes the proof.

The result of Proposition 3.2.4 is valid for any continuous d.f. G, but not for discontinuous G since then (3.19) no longer holds. Using it together with (3.15) enables us to calculate easily the variance function of the o.i. process associated with a renewal process, given the basic interarrival distribution.

3.3. Brownian motion and white noise

Brownian motion (also known as the Wiener process) is the most important and the most fascinating process in probability theory. Its sample paths display a truly astonishing range of behaviour and constitute an absorbing field of study in their own right (the classic reference is [6]). Here we shall do the Brownian motion scant justice by concentrating on its quadratic mean properties, which are comparatively straight-forward. We shall, however, indicate some of the basic sample path properties: the significance of Brownian motion for system theorists is that it provides a well defined mathematical model for certain types of 'noise', particularly in electronics and communication systems (and also, perhaps oddly, for Stock Exchange prices) and one needs to be sufficiently well-informed about its sample behaviour to decide in what ways it can, and in what ways it cannot, be expected to reflect realistically the properties of physical processes.

3.3.1. Definition. *Brownian motion (BM) is a zero-mean process* $\{W_t, t \geq 0\}$ *with stationary normal independent increments. If* $W_0 = 0$ *and* $EW_1^2 = 1$ *then* $\{W_t\}$ *is a* standard *Brownian motion.*

Henceforth Brownian motions will always be taken as standard. According to Proposition 3.1.3 $\{W_t\}$ has covariance function $\sigma^2 t \wedge s$ for some $\sigma^2 (= EW_1^2)$, so that the process $\{(1/\sigma)W_t\}$ is standard. Thus standardization is simply a question of scaling.

Construction of Brownian motion

The finite-dimensional distributions of BM are calculated as follows. Take n times $0 < t_1 < t_2 \ldots < t_n$. Then for $i < j$

$$\text{cov}(W_{t_i}, W_{t_j}) = t_i \wedge t_j = t_i$$

so that the covariance matrix of $Z = (W_{t_i} \ldots W_{t_n})$ is

$$Q = \begin{bmatrix} t_1 & t_1 & t_1 \ldots t_1 \\ t_1 & t_2 & t_2 \ldots t_2 \\ t_1 & t_2 & t_3 & \cdot \\ \cdot & & \cdot & \cdot \\ & & \cdot & \cdot \\ & & \cdot & \cdot \\ t_1 & & & t_n \end{bmatrix}$$

and hence the characteristic function of Z is

$$\phi_z(u) = \exp(-\tfrac{1}{2}u'Qu)$$

It is easily checked that the family of finite dimensional distributions thus defined satisfies the Kolmogorov conditions, so that, yes, Virginia, there *is* a Brownian motion. However, there is a much more direct way of constructing it. Consider first the time interval $[0, 1]$ and let $\{\phi_n, n \in Z^+\}$ be an o.n. basis for $L_2[0, 1]$ (for example, the Haar functions introduced in Proposition 2.2.6). Let $\{\xi_n, n \in Z^+\}$ be an i.i.d. sequence of $N(0, 1)$ r.v.'s, which can be defined, for example, on the probability space $(\Omega, \mathscr{F}, P)$ of Proposition 2.1.5.

3.3.2. Theorem. *For $t \in [0, 1]$ let*

$$W_t^n = \sum_{i=0}^{n} \xi_i \int_0^t \phi_i(s)\,ds \qquad (3.20)$$

Then for each t, $\{W_t^n, n \in Z^+\}$ is a Cauchy sequence in $\mathcal{H}$, and the limiting process is a standard Brownian motion.

Proof. For $s, t \in [0, 1]$ let

$$I_t(s) = \begin{cases} 1 & s < t \\ 0 & s \geqslant t \end{cases} \tag{3.21}$$

Then denoting the inner product in $L_2[0, 1]$ by $\langle .\, ,.\rangle$, we have

$$\int_0^t \phi_i(s)\,\mathrm{d}s = \langle I_t, \phi_i \rangle .$$

Since $\{\phi_i\}$ is an o.n. basis we have

$$I_t = \sum_{i=1}^{\infty} \langle I_t, \phi_i \rangle \phi_i$$

(in the sense of convergence in $L_2[0, 1]$) and from the Parseval equality (Proposition 2.2.5)

$$\|I_t\|^2 = t = \sum_{i=1}^{\infty} \langle I_t, \phi_i \rangle^2 . \tag{3.22}$$

Now from (3.13) we have, for $n > m$,

$$E(W_t^n - W_t^m)^2 = E\left(\sum_{i=m+1}^{n} \xi_i \int_0^t \phi_i(s)\,\mathrm{d}s \right)^2$$

$$= \sum_{i=m+1}^{n} \left(\int_0^t \phi_i(s)\,\mathrm{d}s \right)^2$$

$$= \sum_{i=m+1}^{n} \langle I_t, \phi_i \rangle^2 .$$

But from (3.22) this is a convergent series, so that $E(W_t^n - W_t^m)^2 \to 0$ as $n, m \to \infty$, i.e. $\{W_t^n\}$ is a Cauchy sequence in $\mathcal{H}$; denote its limit by $\{W_t\}$. Again using Proposition 2.2.5,

$$(W_t, W_s) = EW_t W_s = \sum_{i=1}^{\infty} \langle I_t, \phi_i \rangle \langle I_s, \phi_i \rangle$$

$$= \langle I_t, I_s \rangle = t \wedge s .$$

Thus from Proposition 3.1.3 $\{W_t\}$ has stationary orthogonal increments. Also $W_t - W_s \in L(\xi_n, n \in Z^+)$ so that the increments are normal according to Proposition 2.3.7. Thus $\{W_t\}$ is a standard BM.

Note that Theorem 3.3.2 says

$$\mathcal{H}_1^W = \mathcal{L}\{W_t, 0 \leqslant t \leqslant 1\} \subset \mathcal{L}\{\xi_i, i \in Z^+\}.$$

We shall show in Problem 3.5.1 that these subspaces are equal, i.e. it is possible to recover each ξ_i by linear operations on $\{W_t\}$. Thus $\{\xi_i\}$ is an o.n. basis for $\mathcal{H}_1^W$ (which is certainly separable since $\{W_t\}$ is q.m. continuous).

To create a BM on $[0, \infty)$ we piece together a sequence of independent BM's W^1, W^2 ... defined on $[0, 1]$. Indeed, let

$$W_t = W_t^1 \quad \text{for} \quad t \in [0, 1]$$
$$W_t = W_n + W_{t-n}^{n+1} \quad \text{for} \quad t \in (n, n+1].$$

Then it is easily checked that $\{W_t, t \geqslant 0\}$ is a standard BM. At first sight it seems that we need an *array* of independent $N(0, 1)$ r.v.'s to create $\{W_t\}$:

$$\xi_1^1, \xi_2^1, \xi_3^1 \cdots$$
$$\xi_1^2, \xi_2^2 \cdots$$
$$\xi_1^3 \cdots$$

where the nth row forms W^n, but of course these can be combined into a single sequence by counting them off diagonally, so that $\{W_t\}$ is still defined on the 'i.i.d. sequence probability space' of Proposition 2.1.5.

Sample path properties of BM

Referring to (3.20), W_t^n is just a random linear combination of the continuous functions $\psi_i(t) = \int_0^t \phi_i(s)\mathrm{d}s$. So $\{W^n\}$ has continuous sample paths — in fact they are differentiable and

$$\dot{W}_t^n = \sum_{i=0}^{n} \xi_i \phi_i(t).$$

The limiting process $\{W_t\}$ has continuous paths but they are *not* differentiable. To show the path continuity needs a stronger type of convergence than the q.m. convergence considered in Theorem 3.3.2: one has to show that $W_t^n(\omega) \to W_t(\omega)$ for each sample ω, uniformly in $[0, 1]$. Then $\{W_t(\omega), t \in [0, 1]\}$ is the uniform limit of a sequence of continuous functions, and hence is continuous. The argument is quite delicate and the reader is referred to [8] for the details.

If W_t were differentiable then one would expect that $E(\dot{W}_t^n - \dot{W}_t)^2 \to 0$ so that $\dot{W}_t^n$ would be a Cauchy sequence in $\mathcal{H}$. However, it is easy to see that it is not. If we take for $\{\phi_i\}$ the Haar functions, then for any $t \in [0, 1]$, $\phi_i(t)$ is non-zero once in each group and hence

$$\text{var}(\dot{W}_t^n) = \sum_{i=0}^{n} (\phi_i(t))^2 \to \infty \quad \text{as} \quad n \to \infty \qquad (3.23)$$

So $\dot{W}_t^n$ cannot converge to anything in $\mathcal{H}$.

From the above argument it seems that the Brownian path, though continuous, must be very irregular. Some idea of what it looks like is provided by Fig. 3.2, which shows a sample function of W_t^{32} and $\dot{W}_t^{32}$ for the Haar functions $\{\phi_i\}$. A more precise idea is obtained by considering the *quadratic variation* of the Brownian path on $[0, 1]$. This is defined by

$$QV = \lim_{n \to \infty} Q_n$$

where

$$Q_n = \sum_{k=1}^{2^n} (W_{k/2^n} - W_{(k-1)/2^n})^2.$$

The terms of this sum are independent r.v.'s and of course $(W_{k/2^n} - W_{(k-1)/2^n}) \sim N(0, 1/2^n)$. Thus

$$EQ_n = 1.$$

Now if $X \sim N(0, r^2)$ then $EX^4 = 3r^4$, so that

$$\text{var}(X^2) = EX^4 - (EX^2)^2 = 2r^4.$$

Thus

$$\text{var}(Q_n) = E(Q_n - 1)^2 = 2^n \frac{1}{2^{2n}} = \frac{1}{2^n}$$

which shows that $Q_n \to 1$ in $\mathcal{H}$. In general, the quadratic

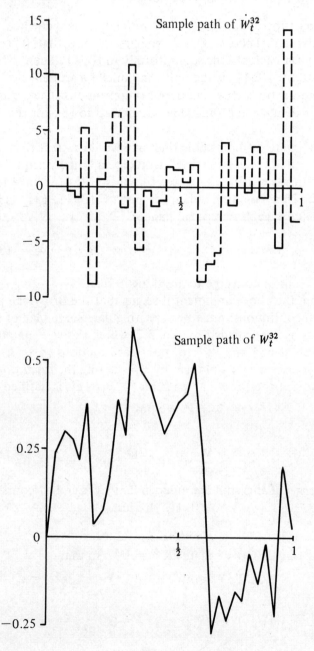

Fig. 3.2

variation on $[0, t]$ is t. The crucial fact is that this convergence also takes place for each sample function (with probability 1), and we can relate this to the idea of path length. For any function $f: [0, 1] \to R$, the *total variation* is

$$TV(f) = \lim_{n \to \infty} \sum_{k=1}^{2^n} \left| f\left(\frac{k}{2^n}\right) - f\left(\frac{k-1}{2^n}\right) \right|.$$

By the triangle inequality the sum on the right increases with n, so that $TV(f)$ is always well-defined but may be $+\infty$. If f is continuous then $TV(f)$ clearly provides a lower bound for the length of the path $\{f(t), 0 \leqslant t \leqslant 1\}$. Now

$$Q_n = \sum_k \left(W\left(\frac{k}{2^n}\right) - W\left(\frac{k-1}{2^n}\right) \right)^2$$

$$\leqslant \left\{ \max_{1 \leqslant k \leqslant 2^n} \left| W\left(\frac{k}{2^n}\right) - W\left(\frac{k-1}{2^n}\right) \right| \right\}$$

$$\sum_k \left| W\left(\frac{k}{2^n}\right) - W\left(\frac{k-1}{2^n}\right) \right|.$$

Thus denoting the first term on the right by m_n we have

$$TV(W) \geqslant \frac{Q_n}{m_n}.$$

Now, as stated above, $Q_n \to 1$, and for any *continuous* function $m_n \to 0$ as $n \to \infty$. Thus $TV(W) = +\infty$, i.e. the Brownian path has infinite length. There was nothing special about the time 1 in the above, so that actually the path has infinite length on any interval $[s, t]$. Note that the argument depends entirely on the sample continuity of $\{W_t\}$. If we took a sample function of a Poisson process, for example, then $m_n \geqslant 1$ for all n (unless the path is identically zero) so the quadratic variation implies nothing about the path length in this case. It is worth remarking in passing that since BM and the Poisson process have the same covariance function, BM is the normal process equivalent to the Poisson process in the sense of Proposition 2.1.8. This shows that the normal equivalent of a process, though having the same quadratic mean properties, may differ radically from the original process as regards sample function behaviour.

We have seen that the Brownian path is a continuous function, but fluctuates extremely rapidly so that it has infinite length and is not differentiable at any point. These are the strengths and drawbacks of BM as a model of a physical process. BM was originally introduced to describe the motion of a small particle suspended in fluid and subject to random molecular bombardment. It seems reasonable to assume that the molecules hitting the particle in one time interval do not affect those doing so at other intervals; hence the independent increments property. Now the path of a physical particle must be continuous, as BM is, but the particle should also have finite velocity, which is inconsistent with BM's infinite path length. As usual in mathematical modelling, BM is a compromise between physical reality and mathematical tractability. The idealization comes in when we assume that the increments over intervals (s, t), $(t + \delta, t + \delta + r)$ are independent even for $\delta = 0$, whereas physically the most that it is reasonable to assert is that these increments are effectively independent for δ bigger than some δ_0 (possibly very small). By taking $\delta_0 = 0$ we get a simpler mathematical model but lose the finite velocity. The caveat is that no properties of the model that depend on the 'fine structure' of the paths, i.e. on what happens in time intervals less than the δ_0 appropriate for the physical problem, can be expected to correspond to anything physically meaningful. As an example of the type of pitfalls that arise, suppose we observe a non-standard BM $\{W_t\}$ and wish to estimate its variance $\sigma^2 = EW_1^2$. Take any $\delta > 0$ and compute the quadratic variation of the sample path $\{W_t, 0 \leqslant t \leqslant \delta\}$. As is easily seen from the preceding argument, this is equal to $\sigma^2 \delta$. Thus apparently, we can estimate the variance *exactly* by observing the process over an arbitrarily short time interval! This is obviously unrealistic, and it is so because the quadratic variation is a 'fine structure' property.

White noise

Physically, the term white noise is used to describe random fluctuations, $\{\zeta_t\}$ arising mainly in electronics and communication systems, which have the following properties
(a) At each t, ζ_t is approximately normally distributed;
(b) ζ_t and ζ_s are effectively uncorrelated for $|t - s| > \delta_0$, where δ_0 is 'small'.

The classic case is 'shot noise' in vacuum tubes, i.e. fluctuations of anode current due to the electron stream not arriving in a completely uniform manner. Since this is the superposition of the effects of a large number of 'independent' electrons the central limit theorem implies (a), and, in (b), 'δ_0 small' means 'much less than $1/W_0$' where W_0 is the highest angular frequency of signals in the system.

As a first attempt at modelling white noise one might consider the normal process $\{X_t\}$ with covariance

$$r_0(t, s) = \begin{cases} 1 & s = t \\ 0 & s \neq t \end{cases}.$$

Thus $EX_t^2 = 1$ and X_t, X_s are independent for $t \neq s$. This satisfies the Kolmogorov consistency conditions and hence can be defined on some probability space $(\Omega, \mathscr{B}, P)$. However, it is not a viable model because its sample paths are so irregular. It is, for example, not q.m. continuous at any point since

$$E(X_t - X_s)^2 = 2 \quad \text{for} \quad t \neq s.$$

One can show, though it is beyond our scope to do so here, that its sample paths cannot be measurable functions (so that one cannot attach a meaning to quantities like $\int_0^1 X_s \, ds$), and furthermore it is 'evanescent' in that all its power is concentrated at infinite frequencies, so that if one passes X_t through a filter with any finite bandwidth then the output will be zero.

The difficulties with r_0 are occasioned by insisting on zero correlation even at immediately adjoining points. One could relax this by taking a correlation function of the form

$$r_\alpha(t, s) = \sigma^2 e^{-\alpha|t-s|}$$

so that (b) will be satisfied for, say, $\alpha = \ln (100\sigma^2)/\delta_0$. r_α is a non-negative definite function so that by Proposition 2.1.7 there is a well-defined normal process $\{Y_t\}$ whose covariance function is r_α. Of course, $\sigma^2 = EY_t^2$. $\{Y_t\}$ is a well-behaved (e.g. q.m. continuous) process, known as the Ornstein–Uhlenbeck process. We shall see below that it can be constructed from Brownian motion.

To obtain an idealization of the Ornstein–Uhlenbeck process we want to consider what happens for small δ_0, i.e. large α. If $\alpha \to \infty$ for fixed σ^2, then we get the covariance function r_0. For other results we have to allow σ^2 to vary as well. The procedure is most easily described in the frequency domain, which is not considered elsewhere in this book, and we refer the reader to [10, Section 3.5] for the full story. For a process $\{X_t\}$ with covariance function $r(t, s)$ which just depends on $|t-s|$, i.e. such that

$$r(t, s) = \tilde{r}(|t-s|)$$

for some function $\tilde{r}$, the *spectral density function* Φ is the Fourier fransform of $\tilde{r}$:

$$\Phi(\lambda) = \int_{-\infty}^{\infty} e^{-i\lambda t} \tilde{r}(t) dt.$$

It turns out that $\Phi(\lambda)$ is the *average power* in the process $\{X_t\}$ at angular frequency λ. In particular, for $r_\alpha(t, s)$ we have

$$\Phi_\alpha(\lambda) = \frac{2\sigma^2/\alpha}{1 + (\lambda^2/\alpha^2)}$$

(see Fig. 3.3). Now suppose we take $\sigma^2 = \frac{1}{2}\alpha$ and let $\alpha \to \infty$. Then for very large α, $\{X_t\}$ has almost uniform distribution of power over all frequencies. In optics this is the characteristic of white light. The limiting spectral density is $\Phi_\infty(\lambda) = 1$. Now formally the inverse Fourier transform of this is the Dirac δ-function, i.e.

$$\int_{-\infty}^{\infty} e^{-i\lambda t} \delta(t) dt = 1$$

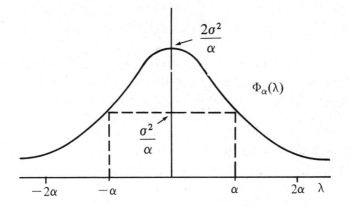

Fig. 3.3

Thus $r_\infty(t, s) = \delta(|t - s|)$. Of course, $\delta(\cdot)$ is not actually a well defined function, but if we did have a process $\{\zeta_t\}$ whose covariance function was r_∞ then formally it would have

(a) $\{\zeta_t\}$ as a normal process
(b) $\text{var}(\zeta_t) = \infty$
(c) $\text{cov}(\zeta_t, \zeta_s) = 0 \quad t \neq s$.

This fictitious process $\{\zeta_t\}$ is called 'white noise'. But we can put it on a rigorous basis as follows. Suppose we define

$$w_t = \int_0^t \zeta_s \, ds.$$

Then for $s < t$ we have

$$\text{cov}(w_t, w_s) = E w_t w_s$$

$$= E \int_0^t \zeta_u \, du \int_0^s \zeta_v \, dv$$

$$= \int_0^s \int_0^t E(\zeta_u \zeta_v) \, du \, dv$$

$$= \int_0^s \int_0^t \delta(u - v)\mathrm{d}u\mathrm{d}v = s.$$

Thus $\mathrm{cov}(w_t, w_s) = t \wedge s$, i.e. $\{w_t\}$ is a standard Brownian motion. So we can formally regard BM as the indefinite integral of white noise. We saw earlier that BM is not in fact differentiable, which explains the difficulties that arise in attempting to set up a white noise model *ab initio*. But we constructed BM as the limit of a sequence $\{W^n\}$ of differentiable processes, so evidently $\{\zeta_t\}$ is related in some way to the 'limit' of $\{\dot{W}^n\}$. However, this is not completely straightforward: although according to (3.23) $\mathrm{var}(W_t^n) \to \infty$, as it should, it is not the case that $\mathrm{cov}(W_t^n, W_s^n) \to 0$ for $s \neq t$, at least for the Haar functions, as is easily checked.

The conclusion we arive at from the above discussion is that we cannot represent mathematically white noise itself, but if it appears in integrated form then BM is an appropriate model. The procedure is best given by an example.

3.3.3. Example. (Estimation of a constant signal in white noise.)

Suppose $\theta \sim N(\mu, \sigma^2)$ and $\{\zeta_t\}$ is a white noise process independent of θ. We observe

$$y_t = \theta + \zeta_t \quad t \in [0, 1] \tag{3.24}$$

What is the best estimate of θ given $\{y_t, 0 \leqslant t \leqslant 1\}$? The problem is not well-posed, but intuitively, since the noise has zero mean and independent values for each t, it seems that the best thing to do would be to 'average it out', i.e. so that the best estimator will be some function of the statistic

$$\int_0^1 y_t \mathrm{d}t = \theta + \int_0^1 \zeta_t \mathrm{d}t.$$

Let us show this. From (3.24) we can write

$$\int_0^t y_s \mathrm{d}s = \theta t + \int_0^t \zeta_s \mathrm{d}s.$$

Now the integral on the right can be replaced by a standard BM $\{W_t\}$, so that, denoting $Y_t = \int_0^t y_s \, ds$, (3.24) becomes

$$Y_t = \theta t + W_t.$$

All the terms here are mathematically well defined, and we can solve the estimation problem easily with the help of Example 1.1.4. Suppose first we observe, not $\{Y_s, s \leqslant 1\}$ but samples $\{Y_{i/n}, i = 1, 2 \ldots n\}$. If we denote $Z_i = Y_{i/n} - Y_{(i-1)/n}$ and $X_i = W_{i/n} - W_{(i-1)/n}$ then

$$Z_i = \frac{1}{n}\theta + X_i$$

and the $\{X_i\}$ are independent, $N(0, 1/n)$. This is the situation considered in Example 1.1.4. Applying the result of Problem 1.2.2 we find that the conditional distribution of θ is $N(\tilde{\mu}, \tilde{\sigma}^2)$, where

$$\tilde{\sigma}^2 = \frac{1}{1 + (1/\sigma^2)}$$

$$\tilde{\mu} = \frac{1}{1 + (1/\sigma^2)} \left(\sum_i Z_i + \frac{\mu}{\sigma^2} \right).$$

But $\sum_i Z_i = Y_1$, so that $\hat{\theta}_n$, the minimum mean-square error estimator of θ, is just

$$\hat{\theta}_n = \tilde{\mu} = \frac{1}{1 + (1/\sigma^2)} \left(Y_1 + \frac{\mu}{\sigma^2} \right).$$

$$= \frac{\sigma^2}{1 + \sigma^2} \tilde{Y}_1 + \mu \qquad (3.25)$$

where $\tilde{Y}_1 = Y_1 - EY_1$ is the centralized observation. Thus θ_n does not depend on n. Now since the processes are normal we know that the best estimate of $(\theta - \mu)$ given $\{Y_s, s \leqslant 1\}$ is just the projection $(\hat{\theta} - \mu)$ onto $\mathcal{L}\{\tilde{Y}_s, s \leqslant 1\}$ and $(\theta_n - \mu) \to (\hat{\theta} - \mu)$ in q.m. since $\{Y_t\}$ is q.m. continuous. Thus $\theta = \hat{\theta}_n$, bearing out our original conjecture that the estimator must be a function of $Y_1 = \int_0^1 y_s \, ds$.

3.4. Wiener Integrals

Let $\{X_t, t \geqslant 0\}$ be a process with stationary orthogonal increments. Scaling if necessary, we can assume that $X_0 = 0$ and $EX_1^2 = 1$ so that the covariance function is $r(t, s) = t \wedge s$ (we will remove the stationarity assumption later). The purpose of this section is to obtain a complete description of the Hilbert subspaces

$$\mathcal{H}_t^X = \mathcal{L}\{X_s, 0 \leqslant s \leqslant t\}$$

and

$$\mathcal{H}^X = \mathcal{L}\{X_s, s \geqslant 0\}.$$

In fact $\mathcal{H}^X$ consists precisely of the set of *Wiener integrals* $\{M(g) : g \in L_2[0, \infty)\}$ of $\{X_t\}$. These are defined as follows.

First consider the set S of *step functions* $g : [0, \infty) \to R$, that is, $g \in S$ if there exist times $0 = t_0 < t_1 \ldots < t_n$ and constants $c_0, c_1 \ldots c_{n-1}$ such that

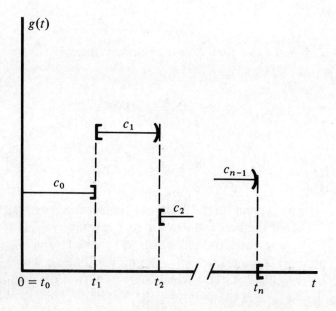

Fig. 3.4

$$g(t) = \begin{cases} c_i & t \in [t_i, t_{i+1}) \quad i = 0, 1 \ldots n-1 \\ 0 & t \geqslant t_n \end{cases} \tag{3.26}$$

(see Fig. 3.4). For $g \in S$ the *Wiener integral of g with respect to $\{X_t\}$* is denoted by $M(g)$ or $\int_0^\infty g(s)\,dX_s$ and defined as:

$$M(g) = \int_0^\infty g(s)\,dX_s = \sum_{i=0}^{n-1} c_i(X_{t_{i+1}} - X_{t_i}). \tag{3.27}$$

The notation $\int g\,dX$ is employed because $M(g)$ is, for step functions, just the Stieltjes integral of g with respect to the sample path of $\{X_t\}$; however, this will not always be true for more general g. As defined by (3.27), $M(g)$ is a r.v. in $\mathcal{H}^X$ and has the following properties

(a) $EM(g) = 0$.

(b) $M(\alpha g + \beta h) = \alpha M(g) + \beta M(h)$ for $g, h \in S$ and $\alpha, \beta \in R$.

(c) $EM(g)M(h) = \int_0^\infty g(s)h(s)\,ds$. \hfill (3.28)

The first of these is evident. For (b) and (c), check that it is no restriction to assume that the t_i's are the same for g and h. Then (b) is clear and, if $d_0 \ldots d_{n-1}$ are the values taken by h,

$$EM(g)M(h) = E\left[\sum_i c_i(X_{t_{i+1}} - X_{t_i})\right]\left[\sum_j d_j(X_{t_{j+1}} - X_{t_j})\right]$$

$$= \sum_{i,j} c_i d_j E(X_{t_{i+1}} - X_{t_i})(X_{t_{j+1}} - X_{t_j})$$

$$= \sum_i c_i d_i (t_{i+1} - t_i)$$

$$= \int_0^\infty g(t)h(t)\,dt.$$

In particular

$$E[M(g)]^2 = \int_0^\infty g^2(t)\,dt$$

Thus $M(g)$ defines an element of $\mathcal{H}^X$ for each $g \in S$, i.e. M can be regarded as a function $M: S \to \mathcal{H}^X$. Now $S \subset L_2[0,\infty)$;

the crucial property of M is that it is inner product preserving, because (3.28c) says precisely

$$(M(g), M(h)) = \langle g, h \rangle$$

where $(.,.)$ is the inner product in $\mathcal{H}^X$ and $\langle .,. \rangle$ in $L_2[0, \infty)$. In view of the linearity (3.28b) this means that *distance* is also preserved, i.e.

$$\| M(g) - M(h) \|_{\mathcal{H}^X} = \| g - h \|_{L_2}.$$

Thus if $\{g_n\}$ is a Cauchy sequence of step functions in L_2, $\{M(g_n)\}$ is a Cauchy sequence in $\mathcal{H}^X$. This enables us to extend the definition of $M(g)$ as follows.

3.4.1. Definition. *Suppose* $g \in L_2$ *has the property that* $g = \lim\limits_n g_n$ *for some sequence* $g_n \in S$. *Then*

$$M(g) = \lim M(g_n).$$

This definition makes sense because $\{M(g_n)\}$ is Cauchy in $\mathcal{H}^X$ and hence has a limit, which clearly cannot depend on the particular sequence $\{g_n\}$ chosen to approximate g. The question now arises as to what class of functions g has the property mentioned in the definition. We already have most of the answer.

3.4.2. Proposition. *S is dense in $L_2[0, \infty)$; i.e. for every g in L_2 there exists a sequence $\{g_n\} \subset S$ such that $\| g - g_n \|_{L_2} \to 0$ as $n \to \infty$.*

Proof. Take $\epsilon > 0$ and choose an integer N such that

$$\| g(1 - I_N) \|^2 = \int\limits_N^\infty g^2(t)\,dt < \frac{1}{4}\epsilon^2$$

(I_N is defined by (3.21)). Now the Haar functions were shown in Proposition 2.2.6 to constitute an o.n. basis for $L_2[0, 1]$ and they can be extended in an obvious manner to form an o.n. basis for $L_2[0, N]$. Thus since $g I_N \in L_2[0, N]$, there exists $h \in L_2[0, N]$ such that $\| g I_N - h \|_{L_2[0,N]} < \epsilon/2$ and h is a finite combination of Haar functions. But then the function h' given by

$$h'(t) = \begin{cases} h(t) & t < N \\ 0 & t > N \end{cases}$$

is a step function and $\|g - h'\|_{L_2[0,\infty)} < \epsilon$. This completes the proof.

The significance of this result is that the Wiener integral $M(g) = \int_0^\infty g(s)\mathrm{d}X_s$ is now defined for all $g \in L_2[0, \infty)$. We also define the *indefinite integral* as follows

$$\int_0^t g(s)\mathrm{d}X_s = M(gI_t)$$

For this we require $gI_t \in L_2[0, \infty)$, i.e. $g \in L_2[0, t]$. Thus $\int_0^t g(s)\mathrm{d}X_s$ is defined for all t if

$$\int_0^t g^2(s)\mathrm{d}s < \infty \quad \text{for all} \quad t.$$

Of course, this does not imply $g \in L_2[0, \infty)$ (think of $g \equiv 1$).

The following result is immediate from the continuity of the Hilbert space inner product.

3.4.3. Proposition. *Properties (3.28) (a), (b) and (c) are satisfied for all $g, h \in L_2[0, \infty)$.*

Here now is the main result.

3.4.4. Theorem. *(a)* $\mathcal{H}^X = \{\int_0^\infty g(s)\mathrm{d}X_s : g \in L_2[0, \infty)\}$
(b) For any $t > 0$,

$$\mathcal{H}_t^X = \left\{\int_0^t g(s)\mathrm{d}X_s : g \in L_2[0, t]\right\}$$

Proof. We will only prove (a) since the argument for (b) is exactly similar. We have already shown that

$$\{M(g) : g \in L_2\} \subset \mathcal{H}^X$$

so it remains to show the reverse inclusion.
If $Y \in \mathcal{H}^X$ then by definition

$$Y = \lim_n Y_n$$

where each Y_n is of the form

$$Y_n = \sum_{i=1}^{k^n} \alpha_i^n X_{t_i^n}.$$

However, we can rearrange this sum as follows (temporarily omitting the superscript n).

$$\sum_{i=1}^{k} \alpha_i X_{t_i} = \alpha_k (X_{t_k} - X_{t_{k-1}})$$
$$+ (\alpha_k + \alpha_{k-1})(X_{t_{k-1}} + X_{t_{k-2}}) + \ldots$$
$$= \sum_{i=1}^{k} \beta_i (X_{t_i} - X_{t_{i-1}})$$

where $t_0 = 0$ and

$$\beta_i = \sum_{j=i}^{k} \alpha_j.$$

Defining the function β^n by

$$\beta^n(t) = \begin{cases} \beta_i & t \in [t_{i-1}, t_i) \\ 0 & t \geqslant t_k \end{cases}$$

we thus have

$$Y_n = \int_0^\infty \beta^n(s)\,\mathrm{d}X_s.$$

Now $\{Y_n\}$ is Cauchy in $\mathcal{H}^X$ so that $\{\beta^n\}$ is Cauchy in $L_2[0,\infty)$ and there exists $\beta \in L_2[0,\infty)$ such that $\beta^n \to \beta$. By Definition 3.4.1

$$Y = \int_0^\infty \beta(s)\,\mathrm{d}X_s$$

i.e. $Y \in \{M(g); g \in L_2\}$. Thus $\mathcal{H}^X \subset \{M(g): g \in L_2\}$ as required.

We introduced above the indefinite integral $\int_0^t g(s)\,\mathrm{d}X_s$; as t varies this forms a process, say $\{Z_t\}$, which is q.m. continuous since

$$E(Z_t - Z_s)^2 = E\int_s^t g(u)\,\mathrm{d}X_u = \int_s^t g^2(u)\,\mathrm{d}u$$

where of course

$$\int_s^t g(u)\,dX_u = M(g(I_t - I_s)).$$

It is not hard to see that $\{Z_t\}$ has orthogonal increments, and in fact all o.i. processes with respect to $\{X_t\}$ arise in this way. Thus Wiener integration is another way in which o.i. processes can be generated. The next result makes these statements precise.

3.4.5. Theorem. *Let $\{X_t\}$ be a process with stationary orthogonal increments and $\{Z_t\}$ another q.m. continuous second-order process. Then the following statements are equivalent.*

(a) For each $t \in R^+, Z_t \in \mathcal{H}_t^X,$ and for $s \leqslant t$

$$(Z_t - Z_s) \perp \mathcal{H}_s^X$$

(b) There exists $g: [0, \infty) \to R$ such that

$$\int_0^t g^2(s)\,ds < \infty \quad \text{for all} \quad t$$

and

$$Z_t = \int_0^t g(s)\,dX_s.$$

Proof. Suppose (b) holds. Then certainly $Z_t \in \mathcal{H}_t^X$, and as remarked above, $Z_t - Z_s = M(g(I_t - I_s))$; hence from Definition 3.4.1,

$$Z_t - Z_s \in L\{(X_v - X_u): s \leqslant u \leqslant v \leqslant t\}.$$

But this subspace is orthogonal to $\mathcal{H}_s^X$, so (a) follows. Note that (a) implies $\{Z_t\}$ has o.i., and that if $\mathcal{P}_s$ is the projection onto $\mathcal{H}_s^X$ then

$$\mathcal{P}_s \left(\int_0^t g(u)\,dX_u \right) = \int_0^s g(u)\,dX_u \qquad (3.29)$$

If (a) holds then for each t there is, according to Theorem 3.4.4, a function $g(t, s)$ such that

$$\int_0^t g^2(t, u)\,du < \infty$$

and

$$Z_t = \int_0^t g(t, u)\,dX_u.$$

We have to show that the functions $\{g(t, .)\}$ are actually the same for all t. Now (a) shows that for $s \leqslant t$

$$\mathscr{P}_s Z_t = \mathscr{P}_s(Z_s + (Z_t - Z_s)) = Z_s.$$

Using (3.29) this gives

$$\int_0^s g(t, u)\,dX_u = \int_0^s g(s, u)\,dX_u,$$

i.e.

$$\int_0^s (g(t, u) - g(s, u))^2\,du = 0.$$

This means that $g(t, .)$ and $g(s, .)$ are equivalent functions in $L_2[0, s]$. It follows from this that if we define, for example

$$g(u) = g(n, u) \quad \text{for} \quad u \in (n - 1, n], n = 1, 2, \ldots$$

then, for any t, $g(.)$ and $g(t, .)$ are equivalent in $L_2[0, t]$, so that g satisfies the conditions stated in (b).

Theorem 3.4.5 will be required in the proof of Theorem 4.4.1 but we can also use it here to extend the idea of Wiener integrals to a class of o.i. processes with non-stationary increments, namely those whose variance functions (see Proposition 3.1.2) are absolutely continuous. Let $\{Z_t\}$ be such a process; then

$$EZ_t^2 = \int_0^t S(u)\,du$$

for some non-negative function S. Now let

$$r(u) = \sqrt{S(u)}$$

and let $\{X_t\}$ be a stationary o.i. process (possibly defined on a different probability space). If we define

$$\tilde{Z}_t = \int_0^t r(u)\,dX_u \tag{3.30}$$

then from Proposition 3.4.3 and Theorem 3.4.4, $\{\tilde{Z}_t\}$ is an o.i. process and $E\tilde{Z}_t^2 = EZ_t^2$. Thus the linear map $f: \mathcal{H}^Z \to \mathcal{H}^{\tilde{Z}}$ which identifies $\tilde{Z}_t$ with Z_t, i.e. defined by

$$f\left(\sum_i \alpha_i Z_{t_i}\right) = \sum_i \alpha_i \tilde{Z}_{t_i}$$

is inner product preserving. Now let Y be an arbitrary element of $\mathcal{H}^Z$ and $\tilde{Y} = f(Y)$. Then $\tilde{Y} \in \mathcal{H}^{\tilde{Z}} \subset \mathcal{H}^X$, so that by Theorem 3.4.4,

$$\tilde{Y} = \int_0^\infty \phi(s)\,\mathrm{d}X_s$$

for some $\phi \in L_2[0, \infty)$. The crucial point is that we can take ϕ to be zero where $r = 0$; formally $\phi = \phi I_{(r>0)}$ in the sense of equality in $L_2[0, \infty)$. Indeed suppose

$$r(t) = 0, \quad t \in [a, b].$$

Then for any $s, t \in [a, b]$, $X_t - X_s \perp \mathcal{H}^{\tilde{Z}}$ because for arbitrary $\tau \in R^+$

$$(X_t - X_s, \tilde{Z}_\tau) = E\int_s^t \mathrm{d}X_\sigma \int_0^\tau r(u)\mathrm{d}X_u = \int_{\tau \wedge s}^{\tau \wedge t} r(u)\mathrm{d}u = 0.$$

But $\tilde{Y} \in \mathcal{H}^{\tilde{Z}}$ by assumption, so

$$0 = E\int_0^\infty \phi\,\mathrm{d}X \int_s^t \mathrm{d}X = \int_s^t \phi(u)\,\mathrm{d}u.$$

Since this holds for all $s, t \in [a, b]$, $\phi(u) = 0, u \in [a, b]$. Now define $\chi: R^+ \to R$ by

$$\chi = \frac{\phi}{r} I_{(r>0)}$$

Then we can write

$$\tilde{Y} = \int_0^\infty \chi(u)r(u)\mathrm{d}X_n$$

$$= \int_0^\infty \chi(u)\mathrm{d}\tilde{Z}_n, \qquad (3.31)$$

the second line coming from (3.30) which we can formally

write $d\tilde{Z} = rdX$. χ has the property that $\phi = \chi r \in L_2[0, \infty)$, i.e.

$$\int_0^\infty \chi^2(u)S(u)du < \infty \qquad (3.32)$$

This just says that χ is square integrable with respect to the measure μ on $[0, \infty)$ given by

$$\mu[a, b) = \int_a^b S(u)du. \qquad (3.33)$$

We have now shown that to every $Y \in \mathcal{H}^Z$ there corresponds a function χ satisfying (3.32); in analogy with (3.31) we write

$$Y = \int_0^\infty \chi(u)dZ_u.$$

It is easily checked that if g is a step function as in (3.26) then

$$\int_0^\infty g(u)dZ_n = \sum_{i=0}^{n-1} c_i(Z_{t_{i+1}} - Z_{t_i})$$

as the notation suggests, and that if g, h satisfy (3.32) then

$$E\int_0^\infty gdZ \int_0^\infty hdZ = \int_0^\infty g(t)h(t)S(t)dt.$$

The above shows that, in the non-stationary case, the Hilbert space $\mathcal{H}^Z$ is in one-to-one correspondence with $L_2(R^+, \mathscr{S}, \mu)$ (μ given by (3.33)) instead of, as in the stationary case, with $L_2(R^+, \mathscr{S}, \lambda)$ (λ = Lebesgue measure). In fact, Wiener integrals can be defined with respect to *any* o.i. process $\{X_t\}$ and the general result is that there is a one-to-one correspondence between $\mathcal{H}^X$ and $L_2(R^+, \mathscr{S}, \mu)$ where μ is determined by

$$\mu([a, b)) = \psi(b) - \psi(a) = E(X_b - X_a)^2$$

(see Proposition 3.1.2). However, to prove this would involve establishing a result similar to Proposition 3.4.2 for $L_2(R^+, \mathscr{S}, \mu)$, whereas for stationary o.i. processes — which are all that are required subsequently — this proposition is almost immediate from the known properties of $L_2[0, 1]$. This was the reason for restricting attention at the outset to the stationary case.

The normal case

Suppose $\{X_t\}$ is a standard Brownian motion and let

$$Z_t = \int_0^t g(s)\mathrm{d}X_s$$

for some g such that $g \in L_2[0, t]$ for all t. Then $\{Z_t\}$ is a normal process since $Z_t \in \mathcal{H}_t^X$. It can be thought of as a BM with a distorted time scale. To see this, suppose for simplicity that $g^2(t) > 0$ for all t, and let

$$\tau(t) = \inf \left\{ s : \int_0^s g^2(u)\mathrm{d}u = t \right\}$$

Evidently this map is one-to-one and

$$\tau^{-1}(s) = \int_0^s g^2(u)\mathrm{d}u.$$

It follows that

$$EZ_{\tau(t)}^2 = \int_0^{\tau(t)} g^2(u)\mathrm{d}u = \tau^{-1}(\tau(t)) = t$$

and hence that if we define

$$W_t = Z_{\tau(t)}$$

then $\{W_t\}$ is a Brownian motion (it has stationary, normal orthogonal increments) and

$$Z_t = W_{\tau^{-1}(t)}. \tag{3.34}$$

Since BM has continuous sample paths, this observation shows among other things that $\{Z_t\}$ has continuous paths also, and that they resemble the Brownian path in having infinite length: from (3.34) the quadratic variation on $[0, t]$ is just $\tau^{-1}(t)$.

Brownian integrals such as Z_t are not just Stieltjes integrals in the sample paths, since these are only defined with respect to functions of bounded variation (see [4], Section 9.1 for a complete treatment). The total variation $TV(f)$ for a function $f: R^+ \to R$ was introduced in Section 3.3. If this is finite then the two functions

$$f^+(t) = \tfrac{1}{2}(f(t) + TV(fI_t))$$
$$f^-(t) = -\tfrac{1}{2}(f(t) - TV(fI_t))$$
(3.35)

are monotone increasing and hence define measures μ^+, μ^- on $(R^+, \mathscr{S})$ given by (2.7). We have $f = f^+ - f^-$ and $f = f^+(-f^-)$ if f is monotone increasing (decreasing). Thus for a measurable function $\phi: R^+ \to R$ we define

$$\int_0^\infty \phi(s)\mathrm{d}f(s) = \int_{R^+} \phi\,\mathrm{d}\mu^+ - \int_{R^+} \phi\,\mathrm{d}\mu^-$$

if both terms are finite.

However, the Brownian path $\{W_t(\omega), t \in R^+\}$ has infinite total variation and no decomposition of the type (3.35) is possible, so the Wiener integral $\int_0^t \phi(s)\mathrm{d}W_s$ cannot in general be defined as a Stieltjes integral in the sample paths, but only in the quadratic mean sense of Definition 3.4.1 (i.e. by considering all sample paths simultaneously).

As an example of the use of Brownian motion integrals, consider the process $\{\tilde{U}_t\}$ defined by

$$\tilde{U}_t = \frac{1}{\sqrt{2\alpha}} \int_0^t e^{-\alpha(t-\tau)}\mathrm{d}W_\tau$$

$$= \frac{1}{\sqrt{2\alpha}} e^{-\alpha t} \int_0^t e^{\alpha\tau}\mathrm{d}W_\tau$$

The covariance of $\{\tilde{U}_t\}$ is

$$E\tilde{U}_t\tilde{U}_{t+s} = \frac{1}{2\alpha} e^{-\alpha t}e^{-\alpha(t+s)} \int_0^t e^{2\alpha\tau}\mathrm{d}\tau$$

$$= e^{-\alpha s}(1 - e^{-2\alpha t})$$

Thus $\{\tilde{U}_t\}$ is a normal process and, for large t, $E\tilde{U}_t\tilde{U}_{t+s} \approx$

$e^{-\alpha s}$, so that $\{\tilde{U}_t\}$ is almost the Ornstein–Uhlenbeck process considered in Section 3.3. We can get the O–U process exactly by taking another $N(0, 1)$ r.v. X_0 independent of $\{W_t\}$ and defining

$$U_t = e^{-\alpha t}X_0 + \tilde{U}_t$$

Then $EU_tU_{t+s} = e^{-\alpha(2t+s)} + E\tilde{U}_t\tilde{U}_{t+s} = e^{-\alpha s}$. We shall see in Chapter 4 that U_t is the solution of a stochastic differential equation with input W_t, and in fact U_t can be regarded as the output of a low pass RC filter when the input is white noise. This is a first illustration of the interrelation between stochastic processes and dynamical systems.

Estimation from o.i. processes

Suppose $\{Y_t\}$ is a process with stationary orthogonal increments. Then the best linear estimate of an associated r.v. X given $\{Y_s, s \leqslant t\}$ is just its projection $\mathscr{P}_tX$ onto $\mathcal{H}_t^Y$. The utility of Theorem 3.4.4 is that enables us to identify exactly what this projection is.

3.4.6. Proposition. *For the r.v. X and o.i. process $\{Y_t\}$ as above, denote $\hat{X}_t = \mathscr{P}_tX$. Then*

$$\hat{X}_t = \int\limits_0^t \left(\frac{\mathrm{d}}{\mathrm{d}s}E[XY_s]\right)\,\mathrm{d}Y_s.$$

Proof. Since $\hat{X}_t \in \mathcal{H}_t^Y$,

$$\hat{X}_t = \int\limits_0^t \phi(s)\mathrm{d}Y_s$$

for some $\phi \in L_2[0, t]$, from Theorem 3.4.4. Also $(X - \hat{X}_t) \perp \mathcal{H}_t^Y$, i.e.

$$(X - \hat{X}_t) \perp \int\limits_0^t \psi(s)\mathrm{d}Y_s$$

for all $\psi \in L_2[0, t]$. This is equivalent to:

$$EX \int_0^t \psi(s)\,\mathrm{d}Y_s = E \int_0^t \phi(s)\,\mathrm{d}Y_s \int_0^t \psi(s)\,\mathrm{d}Y_s$$

$$= \int_0^t \phi(s)\psi(s)\,\mathrm{d}s. \tag{3.36}$$

Now choose $\psi(s) = I_{[0,r]}(s)$ for some $r \leqslant t$. Then

$$E(XY_r) = \int_0^r \phi(s)\,\mathrm{d}s.$$

Thus $E(XY_r)$ must be differentiable in r, and

$$\phi(r) = \frac{\mathrm{d}}{\mathrm{d}r} E(XY_r).$$

An exactly similar argument shows that (with the obvious notation)

$$\hat{X}_\infty = \int_0^\infty \left(\frac{\mathrm{d}}{\mathrm{d}s} E[XY_s] \right) \mathrm{d}Y_s.$$

This implies that $(\mathrm{d}/\mathrm{d}s)E[XY_s] \in L_2[0, \infty)$. These results are of little application by themselves, but are of use mainly in conjunction with the 'innovations process' results of Section 4.3 below.

3.5. Problems and Complements

1. Let $\{W_t\}$ be the BM obtained as the q.m. limit of $\{W_t^n\}$ as in Theorem 3.3.2. Show that if $g \in L_2[0, t]$ then

$$\int_0^t g(s)\,\mathrm{d}W_s = \sum_{i=0}^\infty \xi_i \int_0^t g(s)\phi_i(s)\,\mathrm{d}s.$$

Hence for each k,

$$\xi_k = \int_0^t \phi_k(s)\,\mathrm{d}W_s.$$

Thus $\mathcal{H}_1^W = \mathcal{L}(\xi_i, i \in Z^+)$, as mentioned earlier.

2. (Signal detection [29])

(a) A random variable X has one of two possible density functions, f_0 or f_1. Suppose the hypothesis H_0, that the density is f_0, occurs with probability p and that f_0 is regarded as the conditional density of X given that H_0 is true (with a similar interpretation for f_1 which occurs with probability $1 - p$). We choose a set A, observe the value of X and decide that H_0 is false if $X \in A$ and true if $X \notin A$. Show that the probability of making a mistake is minimized if

$$A = \left\{ x : \frac{f_1(x)}{f_0(x)} \geqslant \frac{p}{1 - p} \right\}.$$

$(f_1(x)/f_0(x))$ is called the *likelihood ratio*.

(b) Suppose we are faced with the radar problem of detecting the presence of a signal (a known function $m(t)$) in white noise; that is we observe $\{Y_t, 0 \leqslant t \leqslant 1\}$ and either the signal is present, i.e.

$$dY_t = m(t)dt + dW_t,$$

or it is absent, i.e.

$$dY_t = dW_t.$$

Here $\{W_t\}$ is a BM. Following the approach used in Example 3.3.3, show that if $m = m_k$, a step function, and observations are taken at a finite number of time points, then the log of the likelihood ratio is

$$l(m_k) = \int\limits_0^1 m_k(s)dY_s - \tfrac{1}{2} \int\limits_0^1 m_k^2(s)ds.$$

Thus if $m \in L_2[0, 1]$ and $\{m_k\}$ is an approximating sequence of step functions, then $l(m_k) \to l(m)$. Since $\int\limits_0^1 m^2(s)ds$ is a constant, this indicates that the signal should be declared to be present if

$$K(Y) = \int\limits_0^1 m(s)dY_s > K_0$$

where K_0 is a 'threshold' level depending on m and the prior probabilities. $K(Y)$ is generated by passing Y through a linear filter with impulse response

$$h(t) = m(1 - t).$$

Such a device is called a 'matched filter'. For a filter with arbitrary impulse response h, we have, with signal present,

$$\int_0^1 \tilde{h}(s)\,dY_s = \int_0^1 \tilde{h}(s)m(s)\,ds + \int_0^1 \tilde{h}(s)\,dW_s$$

where $\tilde{h}(s) = h(1 - s)$. The first term on the right is just the inner product of $\tilde{h}, m$ in $L_2[0, 1]$. Show using the Schwarz inequality that the matched filter maximizes the *signal-to-noise ratio*

$$\frac{(\int_0^1 \tilde{h}(s)m(s)\,ds)^2}{E(\int_0^1 \tilde{h}(s)\,dW_s)^2}.$$

3. The signal detection results of Problem 2(b) can also be obtained by the orthogonal expansion result of Problem 1. Since $\{\phi_i\}$ is an arbitrary o.n. set we can assume $\phi_1 = m$. Then with signal present,

$$Y(t) = (1 + \xi_1) \int_0^t m(s)\,ds + \sum_{i=2}^{\infty} \xi_i \int_0^t \phi_i(s)\,ds.$$

Thus

$$Z_1 = \int_0^1 m(s)\,dY_s = \begin{cases} 1 + \xi_1, & \text{signal present} \\ \xi_1, & \text{signal absent.} \end{cases}$$

$$Z_i = \int_0^1 \phi_i(s)\,dY_s = \xi_i, i \geqslant 2, \quad \text{signal present or absent.}$$

Now $\mathcal{L}\{Y_s, 0 \leqslant s \leqslant 1\} = \mathcal{L}\{Z_i, i = 1, 2 \ldots \}$. The Z_i are independent and normal and only the distribution of Z_1 depends on the signal. It follows that the log likelihood ratio is Z_1.

4. Let $\{X_t\}$ be a renewal process, the distribution of whose interarrival times has density function $f(\cdot)$. Consider the

random telegraph signal

$$S_t = \begin{cases} 0, & X_t \text{ even} \\ 1, & X_t \text{ odd.} \end{cases}$$

Show that $n(t) = ES_t$ satisfies the equation

$$n(t) = G(t) - \int_0^t n(t-s)f(s)\,ds$$

where $G(t) = \int_0^t f(s)\,ds$. Solve this when $\{X_t\}$ is a Poisson process and show that $\lim_{t\to\infty} n(t) = \tfrac{1}{2}$. What about other renewal processes?

Show that the compensating process for $\{S_t\}$ is

$$\bar{A}_t = -\ln\left(\frac{F(S_1)F(S_3)}{F(S_2)F(S_4)} \cdots \frac{F(S_{n-2})}{F(S_{n-1})} F(t-T_{n-1})\right)$$

for $t \in [T_{n-1}, T_n)$, n odd (with an analogous definition when n is even). This means that $\bar{Q}_t = S_t - A_t$ is an o.i. process. Using the method of Proposition 3.2.4 show that $\text{var}(\bar{Q}_t) = m(t) = EX_t$.

CHAPTER FOUR

Estimation in dynamical systems

For our purposes, a dynamical system is a physical process which is represented mathematically by a first-order vector differential equation of the form

$$\dot{x}_t = f(x_t, u_t, v_t). \tag{4.1}$$

Here x_t is the *state* of the system at time t, $\{u_t\}$ is the *control* function which we are free, within certain limits, to choose, and $\{v_t\}$ represents *disturbances*, which are beyond our control and will be represented by a stochastic process. An important special case of (4.1) is the nth order scalar differential equation

$$y_t^{(n)} = f(y_t^{(n-1)} \ldots y_t, u_t, v_t)$$

where $y_t^{(i)} = \mathrm{d}^i y_t/\mathrm{d}t^i$, which can of course be transformed into (4.1) by the substitutions $x_t^1 = y_t, x_t^2 = y_t^{(1)} \ldots x_t^n = y_t^{(n-1)}$, giving

$$\dot{x}_t^1 = x_t^2$$
$$\dot{x}_t^2 = x_t^3$$
$$\dot{x}_t^n = f(x_t^n \ldots x_t^1, u_t, v_t).$$

In this book we only consider linear equations (except in Chapter 6). These have the form

$$\dot{x}_t = A(t)x_t + B(t)u(t) + C(t)v_t \tag{4.2}$$

where A, B, C are now matrices with possibly time-varying but at any rate deterministic components. In addition to representing systems which actually are linear (4.2) frequently arises in representing small deviations of x, u, v in (4.1) about a 'nominal trajectory' $\{\bar{x}_t, \bar{u}_t, \bar{v}_t\}$. Here $\bar{v}_t$ will be the mean disturbance value Ev_t and $\{\bar{x}_t\}$ the desired state trajectory. If f is differentiable in all arguments then denoting $\tilde{x}_t = x_t - \bar{x}_t$ we can write

$$\dot{\tilde{x}} = \left(\frac{\partial f}{\partial x}(\bar{x}, \bar{u}, \bar{v})\right)\tilde{x}_t + \left(\frac{\partial f}{\partial u}(\bar{x}, \bar{u}, \bar{v})\right)\tilde{u}_t + \left(\frac{\partial f}{\partial v}(\bar{x}, \bar{u}, \bar{v})\right)\tilde{v}_t$$
$$+ o(\tilde{x}, \tilde{u}, \tilde{v}).$$

If $|\tilde{x}|$, etc., are small then the last term is small and can be thought of as an additional disturbance. We thus recover (4.2), where A, B, C are just the gradients of f in the three variables, evaluated along the nominal trajectory.

Even for the linear equation (4.2) it is difficult to obtain useful results without making some assumptions about the stochastic process $\{v_t\}$, and we will only consider processes related to the white noise process $\{\zeta_t\}$ introduced in Section 3.3. Of course this leaves us with the problem of how to interpret an equation such as (4.2), since $\{\zeta_t\}$ is not a bona fide process; this question is resolved in Section 4.2 below. As one might exprect, the equation is reformulated in terms of Brownian motion. But then the theory applies to any o.i. process, thus widening again the class of systems we are considering.

The main problem considered in this chapter is how to 'track' the state of a linear dynamical system given noisy observations. Thus for the moment there is no element of control: this is introduced in the next chapter. The central results are really those given in Section 4.3 relating to the so-called 'innovations process' (it is a generalization of the innovations sequence introduced in Chapter 1) which enable the observations to be replaced by an equivalent o.i. process. Then using the results of the previous chapter we can quickly derive the celebrated 'Kalman filter' formulas for recursive estimation; these are given in Section 4.4.

4.1. Multidimensional Processes

An n-vector stochastic process is just a collection of n processes $X = \{x_t^1, x_t^2 \ldots x_t^n, t \in T\}$ all defined on the same probability space $(\Omega, \mathscr{B}, P)$. Thus X can be thought of as a function $X : \Omega \times T \to R^n$ whose value at fixed t is the n-vector r.v.

$$x_t = \begin{bmatrix} x_t^1 \\ \vdots \\ x_t^n \end{bmatrix}.$$

(We shall use the convention of denoting vector processes by lower case letters.) The Hilbert space properties of a vector process $\{x_t\}$ are determined by its *mean*

$$m(t) = Ex_t$$

which is a function $m : T \to R^n$, and its *covariance*

$$R(t, s) = E(x_t - m(t))(x_s - m(s)),$$

which is an $n \times n$ matrix-valued function whose i, jth entry is

$$r_{ij}(t, s) = \text{cov}(x_t^i, x_s^j).$$

An n-vector o.i. process is a vector process $\{z_t\}$ with the property that for each i and $s \leqslant t$

$$(z_t^i - z_s^i) \perp \mathcal{H}_s^z.$$

Note that, since $(z_{t'}^i - z_{s'}^i) \in \mathcal{H}_s^z$ for $s', t' \leqslant s$, each component process $\{z_t^i\}$ has orthogonal increments; but these processes do not necessarily span orthogonal subspaces since $(z_t^i - z_s^i)$ and $(z_{t'}^j - z_{s'}^j)$ may be correlated for overlapping (s, t) and (s', t'). Clearly the covariance matrix $R(t, s)$ of an o.i. process $\{z_t\}$ satisfies

$$R(t, s) = \Gamma(t \wedge s)$$

where $\Gamma(t) = \text{cov}(z_t)$, and $\{z_t\}$ has stationary increments if

$$\Gamma(t) = \Gamma t$$

for some fixed non-negative definite matrix Γ. If Γ is diagonal then the components $\{z_t^i\}$ do span orthogonal subspaces. In particular a vector Brownian motion $\{w_t\}$ is a normal vector process with covariance $I_n t \wedge s$ (I_n is the $n \times n$ identity matrix). This means that $\{w_t\}$ is just a collection of n independent scalar BM's $\{w_t^i\}$.

Now suppose that $\{w_t\}$ is an o.i. process with covariance $I_n t \wedge s$ (but not necessarily normal). For an $m \times n$ matrix-valued function B, the Wiener integral with respect to $\{w_t\}$ is denoted by

$$x_t = \int_0^t B(s)\,dw_s$$

and is an m-vector process with jth component

$$x_t^j = \sum_i \int_0^t b_{ji}(s)\,dw_s^i$$

where b_{ji} is the j, ith element of B. This integral exists as long as all the individual integrals are well defined, which is assured if

$$\sum_{i,j} \int_0^t b_{ji}^2(s)\,ds < \infty \tag{4.3}$$

The salient facts about $\{x_t\}$ are summarized in the following proposition.

4.1.1. Proposition. (a) *If $\{x_t\}$ and $\{y_t\}$ are given by*

$$x_t = \int_0^t B(s)\,dw_s, \quad y_t = \int_0^t C(s)\,dw_s$$

where B, C are $m \times n$ and $r \times n$ matrix-valued functions satisfying conditions similar to (4.3), then

$$Ex_t = Ey_t = 0$$

$$\mathrm{cov}(x_t, y_s) = Ex_t y_s' = \int_0^{t \wedge s} B(u)C'(u)\,du$$

(b) *Theorem 3.4.4 extends to the vector case, i.e. $Z \in \mathcal{H}^w$ if and only if*

$$Z = \int_0^\infty b(s)\,dw_s \tag{4.4}$$

for some $b(s) = (b_1(s) \ldots b_n(s))$ satisfying

$$\sum_i \int_0^\infty b_i^2(s)\,ds < \infty.$$

Remark

Notice in particular that if Γ is an $n \times n$ non-negative definite matrix it can be factored as $\Gamma = DD'$ and then $x_t = \int_0^t D\,dw_s$ has $\text{cov}(x_t) = \Gamma t$ so that any process with stationary o.i. can be generated from mutually orthogonal scalar o.i. processes in this way.

Proof. (a) $Ex_t = Ey_t = 0$ is evident. Now take, say $t > s$. Then $y_s, x_s \in \mathcal{H}_s^w$ and

$$\int_s^t C(u)\,dw_u \perp \mathcal{H}_s^w.$$

Thus

and

$$\text{cov}(x_t, y_s) = \text{cov}(x_s, y_s)$$

$$E(x_s^i y_s^j) = E\left(\sum_{p=1}^n \int_0^s b_{ip}(u)\,dw_u^p\right)\left(\sum_{q=1}^n \int_0^s c_{jq}(u)\,dw_u^q\right)$$

$$= \sum_p \int_0^s b_{ip}(u)c_{jp}(u)\,du$$

$$= \int_0^s (B(u)C'(u))_{ij}\,du.$$

The result follows.

(b) Since $\mathcal{H}^{w^i} \perp \mathcal{H}^{w^j}$ for $i \neq j$, any $Z \in \mathcal{H}^w$ has the unique decomposition

$$Z = Z_1 + Z_2 + \ldots + Z_n$$

where $Z_i \in \mathcal{H}^{w^i}$. But by Theorem 3.4.4, $Z_i \in \mathcal{H}^{w^i}$ if and only if

$$Z_i = \int_0^\infty b_i(s)\,dw_s^i$$

for some b_i satisfying $\int_0^\infty b_i^2(s)\,ds < \infty$ and Z is then given by (4.4) as claimed.

4.2. Linear Stochastic equations

Deterministic Systems

Let us recall first the basic facts about the first order linear vector equation

$$\dot{x} = A(t)x + B(t)u \qquad (4.5)$$

The reader can consult [22] for full details. We suppose that $A(t)$, $B(t)$ are, respectively, $n \times n$ and $n \times r$ matrices whose elements are piecewise-continuous functions of time. Often we will suppress the time dependence and write A, B for $A(t)$, $B(t)$, but no results in the sequel are limited to the time-invariant case, where A and B are constant, unless this is explicitly stated. Consider first the homogeneous matrix equation

$$\dot{X} = AX, X(0) = X_0.$$

For any non-singular $n \times n$ matrix X_0 there is a unique $n \times n$ matrix-valued function $X(t)$ which satisfies this, and furthermore $X(t)$ is non-singular for all t. Now define

$$\Phi(t, s) = X(t)X^{-1}(s).$$

This is the *transition matrix* corresponding to $A(t)$. It has the following properties.

4.2.1. Proposition. (a) $\partial/\partial t\ \Phi(t, s) = A(t)\Phi(t, s)$

$$\Phi(t, t) = I.$$

(b) *Semigroup property*: $\Phi(t, s)\Phi(s, r) = \Phi(t, r)$.
(c) $\Phi(t, s)$ *is non-singular for all t, s.*
(d) $\partial/\partial s\ \Phi(t, s) = -\Phi(t, s)A(s)$.

The first three of these follow immediately from the definition. Notice from (a) that the ith column of $\Phi(t, s)$ is the solution of the linear vector equation $\dot{x} = Ax, x'_s = (0, 0 \ldots 1, 0 \ldots 0)$ (1 in the ith position), and this shows that $\Phi(t, s)p$ is the solution of $\dot{x} = Ax$ with $x_s = p$. Part (d) follows from (b); taking $r = t$ we get

$$\Phi(t, s)\Phi(s, t) = I_n$$

and hence

$$0 = \frac{\partial}{\partial s}(\Phi(t, s)\Phi(s, t))$$

$$= \left(\frac{\partial}{\partial s}\Phi(t, s)\right)\Phi(s, t) + \Phi(t, s)A(s)\Phi(s, t).$$

The result follows since Φ is non-singular.

As is well known the unique solution to the inhomogeneous

equation (4.5) starting at $x_s = p$ is given in terms of the transition matrix by

$$x_t = \Phi(t, s)p + \int_s^t \Phi(t, \tau)B(\tau)u(\tau)d\tau$$

and this formula is valid certainly for all piecewise-continuous r-vector-valued functions $u(\cdot)$.

A basic question which arises in linear system theory is that of *controllability*, i.e. whether it is possible to transfer the system from a given state p at time s to some desired state q at time t, by suitable choice of the 'control' function $\{u(\tau), s \leq \tau \leq t\}$. If so the system is said to be 'controllable on $[s, t]$', and there exists u such that

$$q = \Phi(t, s)p + \int_s^t \Phi(t, \tau)B(\tau)u(\tau)d\tau.$$

But using (b) of Proposition 4.2.1, this is equivalent to

$$0 = \Phi(t, s)[p - \Phi(s, t)q] + \int_s^t \Phi(t, \tau)B(\tau)u(\tau)d\tau$$

which shows that u drives an initial state $[p - \Phi(s, t)q]$ to 0 at time t. Thus for controllability it is enough to require that any initial state can be driven to 0. The following result gives a criterion for controllability. A proof is included since without one it does not seem immediately clear why the condition given should be the appropriate one.

4.2.2. Proposition. *The system (4.5) is controllable on [s, t] if and only if the matrix M defined below is non-singular*

$$M = \int_s^t \Phi(t, \tau)B(\tau)B'(\tau)\Phi'(t, \tau)d\tau \tag{4.6}$$

Proof. Let $G(\tau) = \Phi(t, \tau)B(\tau)$. Then an arbitrary initial state $x_s = p \in R^n$ is steered to 0 by a control $\{u(\tau)\}$ if and only if

$$-\Phi(t, s)p = \int_s^t G(\tau)u(\tau)d\tau.$$

Since $\Phi(t, s)$ is non-singular, this says that we must be able to produce, for arbitrary $y \in R^n$, a function u such that

$$y = \int_s^t G(\tau)u(\tau)\mathrm{d}\tau.$$

It seems clear that this will be possible if and only if y is in the range of $G(\tau)$ for some time interval of positive length; this is equivalent to saying that $\int_s^t |y'G(\tau)|^2\mathrm{d}\tau > 0$. The condition given in the proposition statement just says that this holds for all $y \in R^n$, $y \neq 0$.

Indeed, if $M = \int_s^t GG'\mathrm{d}\tau$ is non-singular then we can write

$$y = MM^{-1}y = \int_s^t G(\tau)G'(\tau)M^{-1}y\,\mathrm{d}\tau.$$

so that the appropriate function u is

$$u(\tau) = G'(\tau)M^{-1}y, \qquad (4.7)$$

whereas if M is singular then there exists $y \in R^n$ such that $|y| > 0$ and $\int_s^t |y'G(\tau)|^2\mathrm{d}\tau = 0$. This implies that $y'G(\tau) = 0$ except at a finite number of points since $G(\cdot)$ is piecewise continuous, and hence that $\int_s^t y'G(\tau)u(\tau)\mathrm{d}\tau = 0$ for any measurable function u. But then

$$y \perp \int_s^t G(\tau)u(\tau)\mathrm{d}\tau$$

for all u, so that the system is not controllable.

Remarks

(a) Notice from (4.7) that if $x_s = p$ can be steered to $x_t = q$ at all, it can be done with a piecewise-continuous function, so that such functions form the 'natural' class of controls. This is related to our initial assumption that $A(\cdot), B(\cdot)$ were piecewise continuous; if they are continuous, then continuous controls suffice.

(b) If A, B are *constant* matrices, then an equivalent criterion for controllability is:

$$\mathrm{rank}\ [B\ \ AB\ \ A^2B\ \ \ldots\ \ A^{n-1}B] = n,$$

i.e. this $n \times nr$ matrix should have n linearly independent columns, see [22, Section 13]. This does not involve the time interval (s, t) so that time-invariant systems are controllable on every interval or on none. Since controllability is just a property of the two matrices A, B one says that (A, B) is a *controllable pair* if the rank condition is satisfied.

Stochastic Systems

We now wish to consider the effect of additive disturbances on the system (4.5). Suppose first that these disturbances form a white noise process as described in Section 3.3. Then (4.5) becomes

$$\dot{x}_t = A(t)x_t + B(t)u(t) + C(t)\zeta_t \qquad (4.8)$$

where $C(t)$ is an $n \times m$ matrix of piecewise continuous functions and $\zeta_t^1 = (\zeta_t^1 \ldots \zeta_t^m)$ is an m-vector of independent white noise. This equation does not make sense as it stands in view of the difficulties in interpreting $\{\zeta_t\}$. But consider the integral version

$$x_t - x_0 = \int_0^t A(s)x_s \, ds + \int_0^t B(s)u(s) \, ds + \int_0^t C(s)\zeta_s \, ds$$

Now if, as in Section 3.3, we regard $\{\zeta_t\}$ as the formal derivative of a Brownian motion process $\{w_t\}$, then we can replace the last term by $\int C \, dw$, thus obtaining the integral equation

$$x_t - x_0 = \int_0^t A(s)x_s \, ds + \int_0^t B(s)u(s) \, ds + \int_0^t C(s) \, dw_s \qquad (4.9)$$

in which all terms are mathematically well defined. Suppose u is a deterministic function (it will not be if we want to consider 'feedback controls'; we return to this point in Chapter 5) and that $\{x_t\}$ is a process which satisfies (4.9). Let $\{x^*(t)\}$ be the solution of

$$\dot{x}^* = Ax^* + Bu, \quad x^*(0) = 0 \qquad (4.10)$$

and define

$$\bar{x}_t = x_t - x^*(t).$$

Then integrating (4.10) we see that $\{x_t\}$ satisfies

$$\bar{x}_t - \bar{x}_0 = \int_0^t A(s)\bar{x}_s \, ds + \int_0^t C(s) \, dw_s. \qquad (4.11)$$

Thus in studying the properties of (4.8) it suffices to consider the control-free equation (4.11). This is customarily called a *stochastic differential equation* and written, in order to be reminiscent of (4.8),

$$d\bar{x}_t = A(t)\bar{x}_t \, dt + C(t) \, dw_t$$

but it should always be borne in mind that the notation is just shorthand for (4.11).

By analogy with the deterministic case, it seems that the solution of (4.8) (with $u \equiv 0$) should be

$$\bar{x}_t = \Phi(t, 0)\bar{x}_0 + \int_0^t \Phi(t, s)C(s)\zeta_s \, ds.$$

Replacing $\zeta_s \, ds$ by dw_s as before gives our candidate for the solution of (4.11), namely

$$\bar{x}_t = \Phi(t, 0)\bar{x}_0 + \int_0^t \Phi(t, s)C(s) \, dw_s.$$

This is in fact correct, as we show below, but we can widen the scope of the discussion by noting that only the second-order properties of the BM are required, and therefore the results apply to any process with stationary orthogonal increments. We need the following technical Lemma.

4.2.3. Lemma. *Let $\{W_t, t \geqslant 0\}$ be a process with stationary orthogonal increments, $g:[0, 1] \times [0, 1] \to R$ be a continuous function and $f, h:[0, 1] \to R$ be measurable functions satisfying*

$$\int_0^1 |f(t)| \, dt < \infty, \qquad \int_0^1 h^2(t) \, dt < \infty$$

Then (a) the process $\{X_t\}$ defined by

$$X_t = \int_0^1 g(t, s)h(s) \, dW_s$$

is q.m. continuous; (b) $A_1 = \int_0^1 f(t)(\int_0^1 g(t, s)h(s)\,dW_s)\,dt$ *and*
$A_2 = \int_0^1 (\int_0^1 f(t)g(t, s)\,dt)h(s)\,dW_s$ *are equivalent random variables in* $\mathcal{H}_1^W$.

Proof. (a) Since g is continuous, it is bounded on $[0, 1]$, so that certainly

$$\int_0^1 g^2(t, s)h^2(s)\,ds < \infty$$

for each t, so that X_t is a well-defined element of $\mathcal{H}_1^W$. Now X_t is defined separately for each t, and to define integrals like $\int_0^1 f(t)X_t\,dt$ we have to be sure that the sample function $\{X_t(\omega), 0 \le t \le 1\}$ is measurable, i.e. we have to consider $\{X_t\}$ simultaneously for all t. According to Proposition 2.3.5, q.m. continuity is a sufficient condition to enable us to piece together the r.v.'s $\{X_t\}$ in a measurable way.

The Hilbert space distance between X_t and $X_{t+\delta}$ is

$$\|X_t - X_{t+\delta}\|^2 = \int_0^1 (g(t, u) - g(t + \delta, u))^2 h^2(u)\,du.$$

Since g is continuous it is uniformly continuous on $[0, 1] \times [0, 1]$, so that given any $\epsilon > 0$ we can choose δ_0 such that

$$|g(t, u) - g(t + \delta, u)| < \epsilon$$

for all $|\delta| < \delta_0$. Thus $\|X_t - X_{t+\delta}\| \to 0$ as $\delta \to 0$. This is (a).

For (b), suppose first that

$$g(t, s) = \sum_{i=1}^n g_{1i}(t)g_{2i}(s)$$

where $\{g_{1i}, g_{2i}, i = 1, 2 \ldots n\}$ are bounded measurable functions. Then it is easily checked that A_1 and A_2 are both equal to

$$\sum_{i=1}^n \int_0^1 f(t)g_{1i}(t)\,dt \int_0^1 g_{2i}(s)h(s)\,dW_s.$$

Now any continuous g can be approximated in this way: given $\epsilon > 0$, choose n such that

$$|g(t, s) - g(t', s')| < \epsilon$$

for $(|t - t'| + |(s - s'|) < 1/n$. Then

$$|g^n(t, s) - g(t, s)| < \epsilon$$

for all $t, s \in [0, 1]$, where

$$g^n(t, s) = \sum_{i=1}^{n} \left(\sum_{j=1}^{n} g\left(\frac{j}{n}, \frac{i}{n}\right) I_{[j/n, j+1/n)}(t) \right) I_{[i/n, i+1/n)}(s).$$

If we now define A_1^n and A_2^n as in (b) but with g replaced by g^n, then $A_1^n = A_2^n$ and it can be verified from Proposition 2.3.6 and Definition 3.4.1 respectively that $A_1^n \to A_1$ and $A_2^n \to A_2$ in q.m. as $n \to \infty$. This completes the proof.

With the lemma in hand, we can prove the main result.

4.2.4. Theorem. *Let $\{w_t, t \geqslant 0\}$ be an m-vector process with stationary orthogonal increments, x an n-vector r.v. orthogonal to $\mathcal{H}^w$, with*

$$m_0 = Ex, Q_0 = \operatorname{cov}(x)$$

and A, C, be matrices of dimension $n \times n$, $n \times m$ respectively whose elements are piecewise continuous functions. Then the stochastic equation

$$dx_t = A(t)x_t dt + C(t)dw_t \tag{4.12}$$
$$x_0 = x$$

has the unique solution

$$x_t = \Phi(t, 0)x + \int_0^t \Phi(t, s)C(s)dw_s \tag{4.13}$$

where Φ is the transition matrix corresponding to A. The moments $m(t) = Ex_t$ and $Q(t) = \operatorname{cov}(x_t)$ are the unique solutions of the equations

$$\dot{m} = Am, m(0) = m_0 \tag{4.14}$$

$$\dot{Q} = AQ + QA' + CC', Q(0) = Q_0. \tag{4.15}$$

Proof. Uniqueness is easily seen: if $\{x_{1t}\}$ and $\{x_{2t}\}$ are two processes satisfying (4.12) then $z_t = x_{1t} - x_{2t}$ satisfies

$$dz_t = A(t)z_t dt, \quad z_0 = 0$$

i.e.
$$\dot{z} = Az, \quad z_0 = 0$$

so that $z_t = 0$ for all t. Thus at most one process satisfies (4.12); our candidate (4.13) is certainly well-defined since (4.3) is satisfied (with $B(s) = \Phi(t, s)C(s)$), so it remains to check that (4.13) satisfies (4.12). With $\{x_t\}$ defined by (4.13) we have

$$
\begin{aligned}
\int_0^t A(s)x_s \, ds &= \int_0^t A(s)\Phi(s, 0)x \, ds \\
&\quad + \int_0^t A(s) \int_0^s \Phi(s, \tau)C(\tau) \, dw_\tau \, ds \\
&= \int_0^t A(s)\Phi(s, 0)x \, ds \\
&\quad + \int_0^t \int_\tau^t A(s)\Phi(s, \tau) \, ds C(\tau) \, dw_\tau \\
&= \int_0^t \frac{\partial}{\partial s} \Phi(s, 0)x \, ds + \int_0^t \int_\tau^t \frac{\partial}{\partial s} \Phi(s, \tau) \, ds C(\tau) \, dw_\tau \\
&= \Phi(t, 0)x - x + \int_0^t (\Phi(t, \tau) - I_n)C(\tau) \, dw_\tau \\
&= x_t - x - \int_0^t C(\tau) \, dw_\tau
\end{aligned}
$$

where we have used Lemma 4.2.3 and Proposition 4.2.1. This establishes the result since (4.12) is to be interpreted as the corresponding integral euqation.

Since the Wiener integral in (4.13) has zero expectation,

$$m(t) = E\Phi(t, 0)x = \Phi(t, 0)m_0$$

from which (4.14) follows. Also, since x and $\mathcal{H}^w$ are orthogonal,

$$Q(t) = \text{cov}(x_t) = Q^1(t) + Q^2(t)$$

where

$$Q^1(t) = \text{cov}(\Phi(t, 0)x)$$

$$Q^2(t) = \text{cov}\left(\int_0^t \Phi(t, s)C(s) \, dw_s\right).$$

From Proposition 4.1.1,

$$Q^2(t) = \int_0^t \Phi(t, s)C(s)C'(s)\Phi'(t, s) \, ds \qquad (4.16)$$

Differentiating this and using Proposition 4.2.1 gives

$$
\begin{aligned}
\dot{Q}^2 &= CC' + \int_0^t \frac{\partial}{\partial t} \Phi(t,s)CC'\Phi'(t,s)\,ds \\
&\quad + \int_0^t \Phi(t,s)CC' \frac{\partial}{\partial t}\Phi'(t,s)\,ds \\
&= CC' + A(t)\int_0^t \Phi CC'\Phi'\,ds + \int_0^t \Phi CC'\Phi'\,ds\,A'(t) \\
&= CC' + AQ^2 + Q^2A'
\end{aligned}
$$

and of course $Q^2(0) = 0$. A similar calculation shows that Q^1 satisfies

$$
\dot{Q}^1 = AQ^1 + Q^1A', \qquad Q^1(0) = Q_0.
$$

(4.15) follows.

Note from (4.13) that $x_s \in \mathcal{L}\{x, w_t, t \leqslant s\}$. Since this subspace is orthogonal to $\mathcal{L}\{(w_u - w_s), u \geqslant s\}$, Theorem 4.2.4 implies that for any $t \geqslant s$,

$$
x_t = \Phi(t,s)x_s + \int_s^t \Phi(t,u)C(u)\,dw_u.
$$

Thus x_t is determined by x_s and $\{(w_u - w_s), s \leqslant u \leqslant t\}$, i.e. x_t retains its role as the 'state' of the system at time t. Evidently this would not be true were $\{w_t\}$ anything other than an o.i. process.

As a first example, let us consider again the Ornstein–Uhlenbeck process, which we can obtain as the solution of the stochastic equation

$$
dU_t = -\alpha U_t dt + c\,dW_t,
$$

where $\{W_t\}$ is a standard BM and α, c are positive constants. Take $EU_0 = 0$; then according to (4.15), $Q(t) = EU_t^2$ is given by

$$
\dot{Q} = -2\alpha Q + c^2.
$$

If we choose $Q_0 = 1$ and $c = \sqrt{2\alpha}$ then $\dot{Q} = 0$ and hence $Q(t) = 1$ for all t. It is easily seen from (4.13) that the general expression for $\mathrm{cov}(x_t, x_s)$ is

$$
\mathrm{cov}(x_t, x_s) = \Phi(t,0)Q_0\Phi'(s,0) + \int_0^{t \wedge s} \Phi(t,u)CC'\Phi'(s,u)\,du.
$$

Applying this in the present case we find $E(U_t U_s)$
$= \exp(-\alpha(t + s - 2(t \wedge s))) = \exp(-\alpha|t - s|)$ in accordance
with the definition in Section 3.3.

This example illustrates the way in which, in stochastic
modelling, the class of additive disturbance processes may be
widened beyond the white noise processes we have considered so far. For suppose, in (4.2), we want to represent
the disturbances v_t by m independent 0–U processes. We
can generate these as the solution of the equation

$$dv_t = \begin{bmatrix} dU_t^1 \\ \vdots \\ dU_t^m \end{bmatrix} = \begin{bmatrix} -\alpha_1 & & 0 \\ & \ddots & \\ 0 & & -\alpha_m \end{bmatrix} \begin{bmatrix} U_t^1 \\ \vdots \\ U_t^m \end{bmatrix} dt + \begin{bmatrix} dW^1 \\ \vdots \\ dW^m \end{bmatrix}$$

Taking $\bar{x}_t' = (v_t', x_t')$ we can now combine this and (4.2) as
follows
$$d\bar{x}_t = \bar{A}\bar{x}_t dt + \bar{B}u(t)dt + \bar{C}dw_t$$
where
$$\bar{A} = \begin{bmatrix} -\alpha_1 & & & \\ & \ddots & & 0 \\ & & -\alpha_m & \\ \hline & C & & A \end{bmatrix}, \quad \bar{B} = \begin{bmatrix} 0 \\ \hline B \end{bmatrix}, \quad \bar{C} = \begin{bmatrix} I_m \\ \hline 0 \end{bmatrix}.$$

This generates the original process $\{x_t\}$ as the last n components of the solution of a stochastic equation of the type
we have already considered. Obviously a variety of additional
effects (e.g. correlation among the components of $\{v_t\}$) can
be introduced by modifying $\bar{A}, \bar{B}, \bar{C}$. In general any equation
of the form (4.2) can be reformulated as a linear stochastic
equation as long as the disturbance process itself can be
modelled that way.

Normal Processes and Controllability

Suppose A, C are constant matrices and $\{w_t\}$ is an m-vector
BM. Then since the equation

$$dx_t = Ax_t dt + Cdw_t, x_0 = 0 \tag{4.17}$$

represents a linear transformation of $\{w_t\}$, $\{x_t\}$ is a normal
process and the r.v. x_t has the characteristic function

$$\phi_t(u) = \exp(im(t) - \tfrac{1}{2}u'Q(t)u)$$

where $m(t)$, $Q(t)$ are given by (4.14) and (4.15) with $m(0) = Q(0) = 0$. There is an interesting tie-up between the properties of these distributions and the controllability of the pair (A, C). Intuitively the idea is the following: controllability of the system

$$\dot{x} = Ax + Cu$$

means that x can be taken anywhere by suitable choice of the control function u. If, in (4.17), we regard $dw_t(= \zeta_t dt)$ as a 'control', then it is being chosen for us at random, and furthermore in a rather irregular way. So it seems that if (A, C) is controllable, $\{x_t\}$ should go everywhere. The precise statement is as follows.

4.2.5. Proposition. *For the system (4.17), the covariance matrix $Q(t) = cov(x_t)$ is non-singular for all $t > 0$ if and only if (A, C) is a controllable pair.*

Proof. One only has to notice that

$$\int_0^t \Phi(t, s)CC'\Phi'(t, s)\,ds$$

is, on the one hand, equal to $Q(t)$ (see (4.16)), and, on the other, is the matrix whose singularity or otherwise determines the controllability of (A, C), according to Proposition 4.2.2.

This result means that if (A, C) is controllable then x_t has the density function

$$f_t(x) = (2\pi \det Q(t))^{-n/2} \exp(-\tfrac{1}{2}x'Q^{-1}(t)x).$$

Now for every $x \in R^n$, $f_t(x) > 0$, and the probability that x_t lies in a small neighbourhood N of x is approximately $f_t(x)V$, where V is the volume of N. Thus we can make the following statement: if N is a set having positive volume and located anywhere in R^n, then there is a strictly positive probability that $x_t \in N$. This is the sense in which $\{x_t\}$ 'goes everywhere'.

Another facet of Proposition 4.2.5 is that, although the result was obtained by quadratic-mean arguments, it certainly implies that the sample paths cannot have bounded velocity. For if the velocity were bounded by some constant K then the time taken to reach the boundary of the sphere

$$S_d = \{x \in R^n : |x| \leqslant d\}$$

would be at least d/K and consequently we would have $P[x_t \in R^n - S_d] = 0$ for $t < d/K$, whereas in fact this probability is positive (though, of course, very small for small t).

Two examples will perhaps clarify the content of the preceding discussion. Let $\{W_t\}$ be a standard BM and consider first

$$A = \begin{bmatrix} 0 & 1 \\ a_1 & a_2 \end{bmatrix} \quad C = \begin{bmatrix} 0 \\ 1 \end{bmatrix}$$

This is a controllable pair since

$$[C, AC] = \begin{bmatrix} 0 & 1 \\ 1 & a_2 \end{bmatrix}$$

and written out in components the corresponding system equations are

$$\begin{aligned} dx_t^1 &= x_t^2 dt, & x_0^1 &= 0 \\ dx_t^2 &= (a_1 x_t^1 + a_2 x_t^2) dt + dW_t, & x_0^2 &= 0 \end{aligned} \qquad (4.18)$$

Thus (x_t^1, x_t^2) has non-singular covariance even though the 'noise' does not enter the x^1 equation directly. Referring back to the discussion at the beginning of this chapter, (4.18) is evidently the stochastic model for a second-order differential equation with white noise disturbances:

$$\ddot{x}_t = a_1 x_t + a_2 \dot{x}_t + \zeta_t.$$

It will be seen in Problem 4.5.3 that the general nth order differential equation behaves similarly.

As a second example, consider

$$A = \begin{bmatrix} 3 & 1 \\ -1 & 1 \end{bmatrix}, \quad C = \begin{bmatrix} 1 \\ -1 \end{bmatrix}$$

In this case

$$[C, AC] = \begin{bmatrix} 1 & 2 \\ -1 & -2 \end{bmatrix}$$

so the corresponding process (x_t^1, x_t^2) must have singular covariance. It is easy to see why. In components we have

$$dx_t^1 = (3x_t^1 + x_t^2)dt + dW_t, \quad x_0^1 = 0$$
$$dx_t^2 = (-x_t^1 + x_t^2)dt - dW_t \quad x_0^2 = 0$$

and hence

$$d(x_t^1 + x_t^2) = 2(x_t^1 + x_t^2)dt$$

Thus $Z_t^1 = x_t^1 + x_t^2 = 0$ for all t and $Z_t^2 = x_t^1 - x_t^2$ satisfies

$$dZ_t^2 = 2(Z_t^2 dt + dW).$$

So in this case there is no noise in the direction of the line $x^1 = x^2$ and the distribution of (x_t^1, x_t^2) is concentrated on the line $x^1 = -x^2$. The distribution along this line is the one-dimensional normal distribution. In terms of Kalman's 'canonical structure theorem' ([24], Section 5.2), Z^1 is the 'uncontrollable part' of the system — the part the input never reaches.

4.3. The Innovations Process

The innovations process plays a central role in estimation theory for dynamical systems. We saw in Chapter 1 that for observations of the form

$$Y_k = X + W_k \quad (4.19)$$

($\{W_k\}$ an i.i.d. sequence) it was possible successively to orthogonalize the $\{Y_k\}$ sequence, producing the 'innovation projections' $\{Z_k\}$ which form an i.i.d. sequence spanning the same subspaces as the original $\{Y_k\}$ and from which recursive estimates of X are easily obtained. In this section we derive the continuous-time analogues of these results. Suppose we have a 'signal' process $\{z_t\}$ and an o.i. process $\{w_t\}$ and we observe

$$y_t = \int_0^t z_s \, ds + w_t$$

Notice in passing that if $\{w_t\}$ is BM then this represents the observations

$$\dot{y}_t = z_t + \zeta_t$$

where $\{\zeta_t\}$ is white noise — the continuous-time version of (4.19). We shall show that there is an o.i. process $\{v_t\}$ which spans the same family of linear subspaces as $\{y_t\}$. In the terminology of Section 3.1: $\{y_t\}$ has multiplicity n. This is a very striking result and its utility is obvious: we can use the information obtained in Chapter 3, about the structure of subspaces generated by o.i. processes, to calculate estimates of the signal $\{z_t\}$.

Let $\{z_t, t \geqslant 0\}$ and $\{w_t, t \geqslant 0\}$ be n-vector and r-vector processes respectively and let $G(\cdot)$ be an $n \times r$ matrix whose elements are piecewise-continuous functions of time.

The following conditions will be required:

(A) $\{w_t\}$ has stationary orthogonal increments.

(B) $\{z_t\}$ is a second-order, q.m. continuous process.

(C) For every s and $t > s$

$$(w_t - w_s) \perp \mathcal{H}_s^{w,z} = \mathcal{L}\{w_\tau, z_\tau, \tau \leqslant s\}.$$

A stronger condition than (C) is:

(C') $\mathcal{H}^w \perp \mathcal{H}^z$

(D) For all $t \geqslant 0$, $G(t)G'(t)$ is strictly positive definite.

In (C'), as usual,

$$\mathcal{H}_t^w = \mathcal{L}\{w_s, s \leqslant t\}, \qquad \mathcal{H}^w = \mathcal{L}\{w_s, s \geqslant 0\}$$

Also note that (D) cannot be satisfied unless $r \geqslant n$. Now define

$$y_t = \int_0^t z_s \, ds + \int_0^t G(s) \, dw_s \tag{4.20}$$

and let $\mathscr{P}_t^y$ be the projection operator onto $\mathcal{H}_t^y$. Let

and
$$\hat{z}_t = \mathscr{P}_t^y z_t$$
$$\tilde{z}_t = z_t - \hat{z}_t.$$

Then the *innovations process* is

$$\nu_t = y_t - \int_0^t \hat{z}_s \, \mathrm{d}s. \qquad (4.21)$$

In making this definition we run into the by now familiar problem that $\hat{z}_t$ is defined for each t but $\hat{z}_t(\omega)$ must be a measurable function of t in order to define $\int \hat{z}_s \, \mathrm{d}s$. In this case $\{y_t\}$ is evidently q.m. continuous so that by the remarks following Proposition 2.3.4, $\mathscr{P}_s^y z_t \to \mathscr{P}_t^y z_t$ as $s \to t$. Also clearly $\mathscr{P}_s^y z_t \to \mathscr{P}_s^y z_s$ as $s \to t$. Hence

$$\hat{z}_t - \hat{z}_s = (\mathscr{P}_t^y z_t - \mathscr{P}_s^y z_t) + (\mathscr{P}_s^y z_t - \mathscr{P}_s^y z_s)$$
$$\to 0 \quad \text{as} \quad s \to t$$

This shows that $\{\hat{z}_t\}$ is q.m. continuous, so that a measurable version exists by Proposition 2.3.5.

Under conditions (A)–(D) the innovations process has orthogonal increments and spans the same family of subspaces as the observations process $\{y_t\}$, i.e. $\mathcal{H}_t^\nu = \mathcal{H}_t^y$ for each t. We prove the latter assertion first under the stronger condition (C') since this results in some simplification and is all that is required in the sequel. However, condition (C) arises naturally if there is feedback from $\{y_t\}$ to $\{z_t\}$, so that z_t depends partly on the past $\{y_s, s \leqslant t\}$ (and hence on $\{w_s, s \leqslant t\}$). The argument for this case is given at the end of the section. $\{\nu_t\}$ is often written in differential form as

$$\mathrm{d}\nu_t = \mathrm{d}y_t - \hat{z}_t \mathrm{d}t$$

or, equivalently

$$\mathrm{d}\nu_t = \tilde{z}_t \mathrm{d}t + G(t)\mathrm{d}w_t. \qquad (4.22)$$

4.3.1. Theorem. *Let conditions (A)–(D) hold. Then (a) for all s and $t > s$,*
$$\nu_t - \nu_s \perp \mathcal{H}_s^y$$
(b) $\operatorname{cov}(\nu_t) = \int_0^t G(s)G'(s)\,ds$

Remark

Since by definition $\nu_s \in \mathcal{H}_s^y$ for each s, (a) implies that $\{\nu_t\}$ is an o.i. process, and (b) says that its covariance is the same as that of the o.i. process $d\xi_t = G(t)dw_t$ in (4.20) or (4.22)

Proof. Take any $X \in \mathcal{H}_s^y$. Denoting the ith component of ν_t by ν_t^i and the elements of G by g^{ij}, we have

$$(X, \nu_t^i - \nu_s^i) = E\left[X\left(\int_s^t \tilde{z}_u^i \, du + \sum_j \int_s^t g^{ij}(u)dw_u^j\right)\right]$$

$$= E\left[X \int_s^t \tilde{z}_u^i \, du\right] + \sum_j E\left[X \int_s^t g^{ij}(u)dw_u^j\right].$$

The last term is 0 from (C), and, since $\tilde{z}_u^i \perp \mathcal{H}_u^y \supset \mathcal{H}_s^y$,

$$E\left[X \int_s^t \tilde{z}_u^i \, du\right] = \int_s^t E[X\tilde{z}_u^i)du = 0$$

Thus $(\nu_t^i - \nu_s^i) \perp \mathcal{H}_s^y$ so that each component of $\{\nu_t\}$ is an o.i. process. Take any partition $s = t_1 < t_2 \ldots < t_n = t$ of $[s, t]$; then

$$E(\nu_t^i - \nu_s^i)^2 = \sum_k E(\nu_{t_{k+1}}^i - \nu_{t_k}^i)^2$$

$$= \sum_k E\left[\int_{t_k}^{t_{k+1}} \tilde{z}_u^i \, du + \sum_j \int_{t_k}^{t_{k+1}} g^{ij}(u)dw_u^j\right]^2$$

$$= \sum_k E\left[\int_{t_k}^{t_{k+1}} \tilde{z}_u^i \, du\right]^2 + \sum_{j,k} E\left[\int_{t_k}^{t_{k+1}} g^{ij}(u)dw_u^j\right]^2$$

$$+ 2\sum_{k,j} E\left[\int_{t_k}^{t_{k+1}} \tilde{z}_u^i \, du \int_{t_k}^{t_{k+1}} g^{ij}(u)dw_u^j\right]. \quad (4.23)$$

Now the second term is just

$$\sum_j \int_s^t (g^{ij}(u))^2 du$$

so to get (*b*) we have to show that the other terms converge to 0 as $\max_k (t_{k+1} - t_k) \to 0$. For the first term

$$E\left(\int_{t_k}^{t_{k+1}} \tilde{z}_u^i \, du\right)^2 = E\left(\int_{t_k}^{t_{k+1}} \tilde{z}_u^i \, du\right)\left(\int_{t_k}^{t_{k+1}} \tilde{z}_v^i \, dv\right)$$

$$= \int_{t_k}^{t_{k+1}} \int_{t_k}^{t_{k+1}} E(\tilde{z}_u^i \tilde{z}_v^i) \, du \, dv$$

$$= \int_{t_k}^{t_{k+1}} \int_{t_k}^{t_{k+1}} \tilde{r}(u, v) \, du \, dv$$

$$\leqslant \int_{t_k}^{t_{k+1}} \int_{t_k}^{t_{k+1}} \sqrt{\tilde{r}(u, u) \tilde{r}(v, v)} \, du \, dv \quad (4.24)$$

by the Schwarz inequality, where

$$\tilde{r}(u, v) = \text{cov}(\tilde{z}_u^i, \tilde{z}_v^i).$$

Now $\tilde{r}(u, u) = \|\tilde{z}_u^i\|^2$ is bounded on $[s, t]$ since $\{z_u^i\}$ is q.m. continuous and $\|\tilde{z}_u^i\|^2 \leqslant \|z_u^i\|^2$. It follows from (4.24) that

$$E\left(\int_{t_k}^{t_{k+1}} \tilde{z}_u^i \, du\right)^2 \leqslant K(t_{k+1} - t_k)^2$$

for some constant K, so that

$$\sum_k E\left(\int_{t_k}^{t_{k+1}} \tilde{z}_u^i \, du\right)^2 \leqslant K \sum_k (t_{k+1} - t_k)(t_{k+1} - t_k)$$

$$\leqslant K(t - s) \max_k (t_{k+1} - t_k)$$

$$\to 0 \text{ as } \max_k (t_{k+1} - t_k) \to 0.$$

For the third term in (4.23) the Schwarz inequality gives

$$\left| E\left[\left(\int_{t_k}^{t_{k+1}} \tilde{z}_u^i \, du\right)\left(\int_{t_k}^{t_{k+1}} g^{ij}(u) \, dw_u^j\right)\right] \right|$$

$$\leqslant \left(E\left[\int_{t_k}^{t_{k+1}} \tilde{z}_u^i \, du\right]^2\right)^{1/2} \left(\int_{t_k}^{t_{k+1}} (g^{ij}(u))^2 \, du\right)^{1/2}$$

$$\leqslant K(t_{k+1} - t_k) \cdot (t_{k+1} - t_k)^{1/2}.$$

Thus the third term in (4.23) is dominated by

$$K \sum_k (t_{k+1} - t_k)(t_{k+1} - t_k)^{1/2} \leqslant K(t-s) \max_k (t_{k+1} - t_k)^{1/2}$$

which also converges to 0 as $\max_k (t_{k+1} - t_k) \to 0$. For different components ν_t^i and ν_t^l an exactly similar argument shows that

$$E(\nu_t^i - \nu_s^i)(\nu_t^l - \nu_s^l) = \sum_j \int_s^t g^{ij}(u)g^{lj}(u) \, du$$

whereas for non-overlapping intervals (s', t'), (s, t), with, say, $t' \leqslant s$, part (a) shows that $(\nu_{t'}^i - \nu_{s'}^i) \perp (\nu_t^l - \nu_s^l)$ since $(\nu_{t'}^i - \nu_{s'}^i) \in \mathcal{H}_s^y$. This completes the proof.

We have now shown that $\{\nu_t\}$ is an o.i. process. To show that it spans the same subspaces as $\{y_t\}$, two preliminary lemmas are required. The first of these concerns the structure of the subspace $\mathcal{H}_t^y$. If y were an o.i. process then all its elements would be Wiener integrals of the form $\int_0^t \beta(s) \, dy_s$. In fact this representation still holds even though $\{y_t\}$ is not an o.i. process, because it has an additive o.i. component. As mentioned earlier, it is convenient at this point to impose condition (C'), i.e. to assume that $\mathcal{H}^z$ and $\mathcal{H}^w$ are orthogonal subspaces. The definition of $\int_0^t \beta(s) \, dy_s$ is just

$$\int_0^t \beta(s) \, dy_s = \int_0^t \beta(s) z_s \, ds + \int_0^t \beta(s) G(s) \, dw_s,$$

each of the terms on the right being well-defined as a q.m. integral.

4.3.2. Lemma. *Suppose conditions (A), (B), (C'), (D) hold and that $X \in \mathcal{H}_t^y$. Then there exist $\beta^1, \beta^2, \ldots, \beta^n \in L_2[0, t]$ such that*

$$X = \sum_i \int_0^t \beta^i(s) \, dy_s^i$$

Proof. Since $X \in \mathcal{H}_t^y$ we can write

$$X = \lim_m X_m$$

where each X_m is a finite linear combination

$$X_m = \sum_i \left(\sum_j \alpha_m^{ij} y_{t_j}^{i\,m} \right)$$

for some constants $\{\alpha_m^{ij}\}$. As in the proof of Theorem 3.4.4 this sum can be rearranged so that

$$X_m = \sum_i \left(\sum_j \beta_m^{ij} (y_{t_{j+1}^m} - y_{t_j^m}) \right).$$

Defining the function $\beta_m : [0, t] \to R^n$ by

$$\beta_m^i (s) = \beta_m^{ij}, \quad s \in [t_j^m, t_{j+1}^m),$$

we then have

$$X_m = \int_0^t \beta_m'(s)\,dy_s$$

$$= \int_0^t \beta_m'(s)z_s\,ds + \int_0^t \beta_m'(s)G(s)\,dw_s.$$

Thus

$$E(X_m - X_p)^2 = E\left[\int_0^t (\beta_m - \beta_p)'z_s\,ds + \int_0^t (\beta_m - \beta_p)'G\,dw_s \right]^2.$$

Denoting the first term on the right by Z and recalling from (C') that $Z \perp \mathcal{H}_t^w$, we get

$$E(X_m - X_p)^2 = E\left(Z + \int_0^t (\beta_m - \beta_p)'G\,dw_s \right)^2$$

$$\geq E\left(\int_0^t (\beta_m - \beta_p)'G\,dw_s \right)^2$$

$$= \int_0^t (\beta_m - \beta_p)'GG'(\beta_m - \beta_p)\,ds.$$

Now since by assumption (D), GG' is strictly positive definite and piecewise continuous, there exists a constant b such that

$G(s)G'(s) \geqslant bI_n$ (i.e. $a'GG'a \geqslant b|a|^2$ for all $a \in R^n$) for all $s \in [0, t]$. Thus

$$E(X_m - X_p)^2 \geqslant \int_0^t (\beta_m - \beta_p)'GG'(\beta_m - \beta_p)\,ds$$

$$\geqslant b \int_0^t (\beta_m - \beta_p)'(\beta_m - \beta_p)\,ds$$

$$= b \sum_i \int_0^t (\beta_m^i - \beta_p^i)^2\,ds$$

But by assumption $E(X_m - X_p)^2 \to 0$ as $m, p \to \infty$; hence, for each i, $\{\beta_m^i\}$ is Cauchy in $L_2[0, t]$ and there exists a limit $\beta^i \in L_2[0, t]$. Now let $\beta(s) = (\beta^1(s) \ldots \beta^n(s))$ and define

$$\bar{X} = \int_0^t \beta(s)z_s\,ds + \int_0^t \beta(s)G(s)\,dw_s$$

Then

$$E(X_m - \bar{X})^2 = E\left[\int_0^t (\beta_m - \beta)z_s\,ds + \int_0^t (\beta_m - \beta)G\,dw \right]^2$$

$$= E\left[\int_0^t (\beta_m - \beta)z_s\,ds \right]^2$$

$$+ \int_0^t (\beta_m - \beta)GG'(\beta_m - \beta)'\,ds$$

Using the q.m. continuity of $\{z_s\}$ and assumption (D), it is easily seen that the terms on the right converge to 0 as $m \to \infty$. This shows that $\bar{X} = X$ and completes the proof.

The next fact has nothing to do with stochastic processes, but is really a result from functional analysis. Equation (4.25) below is a *Volterra integral equation*. There is an extensive theory surrounding these equations, for details of which the reader is referred to [23] or [28], but the result required here is easily proved directly.

We denote by $L_2^n[a, b]$ the set of functions $f : [a, b] \to R^n$ such that each component f^i is in $L_2[a, b]$.

4.3.3. Lemma. *Suppose* $h(t, s) = [h^{ij}(t, s)]$ *is an* $n \times n$ *matrix for each* $t, s \in [a, b]$, *whose elements are measurable functions satisfying*

$$\sum_{i,j} \int_b^a \int_b^a (h^{ij}(t, s))^2 \, ds \, dt < \infty$$

Then for any function $e \in L_2^n[a, b]$ *there is a unique* $\phi \in L_2^n[a, b]$ *satisfying:*

$$\phi(t) = e(t) + \int_a^t h(t, s)\phi(s) \, ds \qquad (4.25)$$

Remarks

(a) Notice from (4.25) that only the values of $h(t, s)$ for $s \leqslant t$ are used, so it may as well be assumed that $h(t, s) = 0$ for $s > t$.

(b) The equality in (4.25) is to be interpreted in the sense that the ith components of the left and right hand sides are equivalent functions in $L_2[a, b]$. The same applies to all similar statements below.

Proof. The following proof is for the one-dimensional case. The general n-dimensional case parallels it exactly but is notationally a good deal more cumbersome.

The idea is to solve (4.25) by the well-known Picard iteration technique as used in connection with differential equations. That is, we define inductively

$$\phi_0(t) = e(t)$$

$$\phi_{m+1}(t) = e(t) + \int_a^t h(t, s)\phi_m(s) \, ds \qquad (m = 0, 1, \ldots)$$

and show that ϕ_m converges in $L_2[a, b]$ to a function ϕ

satisfying (4.25). Now

$$\phi_{m+1}(t) - \phi_m(t) = \int_a^t h(t, s)(\phi_m(s) - \phi_{m-1}(s)) \, ds.$$

Using the Schwarz inequality, we get

$$(\phi_{m+1}(t) - \phi_m(t))^2 \leqslant \int_a^t h^2(t, s) \, ds \int_a^t (\phi_m(s) - \phi_{m-1}(s))^2 \, ds \tag{4.26}$$

In particular, for $m = 1$,

$$(\phi_2(t) - \phi_1(t))^2 \leqslant \int_a^t h^2(t, s) \, ds \int_a^t \int_a^s h^2(s, u) \, du \, ds \int_a^t e^2(s) \, ds \tag{4.27}$$

Now define

$$K = \int_a^b e^2(s) \, ds, \quad D(t) = \int_a^t \int_a^s h^2(s, u) \, du \, ds.$$

Then

$$\dot{D}(t) = \frac{d}{dt} D(t) = \int_a^t h^2(t, u) \, du.$$

We show by induction that

$$(\phi_{m+1}(t) - \phi_m(t))^2 \leqslant \frac{1}{m!} \dot{D}(t) D^m(t) K \tag{4.28}$$

From (4.27) this is true for $m = 1$. Suppose it holds for $m = k$. Then using (4.26)

$$(\phi_{k+2}(t) - \phi_{k+1}(t))^2 \leqslant \dot{D}(t) \frac{1}{k!} \int_a^t \dot{D}(s) D^k(s) \, ds \, K$$

$$= \dot{D}(t) \frac{1}{k!} \int_a^t \frac{d}{ds} \left(\frac{1}{k+1} D^{k+1}(s) \right) ds \, K$$

$$= \frac{1}{(k+1)!} \dot{D}(t) D^{k+1}(t) K$$

Thus (4.28) holds for all m. Integrating both sides,

$$\|\phi_{m+1} - \phi_m\|^2 = \int_a^b (\phi_{m+1}(t) - \phi_m(t))^2 \, dt \leqslant \frac{1}{(m+1)!} D^{m+1}(b) K$$

and hence, using also the inequality $(\Sigma a_n b_n)^2 \leqslant \Sigma a_n^2 \Sigma b_n^2$,

$$\|\phi_{m+k} - \phi_m\|^2 \leqslant \left(\sum_{p=m+1}^{m+k} \frac{2^p}{2^p} \|\phi_p - \phi_{p-1}\| \right)^2 \leqslant \sum_{p=m+1}^{m+k} \frac{D^p(b)}{p!} 2^{2p} K$$

This shows that $\{\phi_m\}$ is a Cauchy sequence in $L_2[a, b]$ since the sum on the right is convergent. Let $\phi = \lim_m \phi_m$. To show that ϕ satisfies (4.25) define

$$\psi(t) = \phi(t) - e(t) - \int_a^t h(t, s)\phi(s)\,ds$$

Then using the recurrence relation for ϕ_m we have

$$\psi(t) = (\phi(t) - \phi_{m+1}(t)) - \int_a^t h(t, s)(\phi(s) - \phi_m(s))\,ds.$$

But each term on the right converges to 0 in $L_2[a, b]$ as $m \to \infty$. Thus $\psi = 0$ and (4.25) is satisfied. A similar argument shows that ϕ is the only function in $L_2[a, b]$ satisfying (4.25).

Armed with the two lemmas, we can now prove that $\{v_t\}$ and $\{y_t\}$ generate the same family of subspaces.

4.3.4. Theorem. *Suppose conditions (A), (B), (C') and (D) hold. Then for each* t, $\mathcal{H}_t^y = \mathcal{H}_t^v$.

Proof. $t > 0$ is fixed throughout the following argument. First, by definition, $v_s \in \mathcal{H}_s^y \subset \mathcal{H}_t^y$ for any $s \leqslant t$, so that $\mathcal{H}_t^v \subset \mathcal{H}_t^y$.

For any $s \in [0, t]$, $\hat{z}_s^i \in \mathcal{H}_s^y$ so that by Lemma 4.3.2 there exist functions $\beta^{ij}(s, u)$ such that

$$\sum_{i,j} \int_0^s (\beta^{ij}(s, u))^2 du < \infty$$

and

$$\hat{z}_s = \int_0^s \beta(s, u)\,dy_u$$

where β is the matrix with elements β^{ij}. Using the q.m. continuity of $\{z_s\}$ one can show that $\beta^{ij}(s, u)$ can be chosen jointly measurable in (s, u). By the same kind of argument as in the proof of Lemma 4.3.2 we find that

$$E|z_s|^2 \geqslant \sum_{i,j} \int_0^s (\beta^{ij}(s, u))^2 du.$$

But $E|z_s|^2$ is bounded on $[0, t]$ since $\{z_s\}$ is q.m. continuous, and hence

$$\sum_{i,j} \int_0^t \int_0^s (\beta^{ij}(s, u))^2 du\,ds < \infty \qquad (4.29)$$

Now for any $f \in L_2^n[0, t]$ we have from (4.21)

$$\int_0^t f'(s)dv_s = \int_0^t f'(u)dy_u - \int_0^t f'(s)\hat{z}_s\,ds$$

$$= \int_0^t f'(u)dy_u - \int_0^t f'(s) \int_0^s \beta(s, u)dy_u\,ds.$$

By Lemma 4.2.3 the order of integration in the last term can be interchanged, and we get

$$\int_0^t f'(s)dv_s = \int_0^t \left[f(u) - \int_u^t \beta(s, u)f(s)ds \right]' dy_u$$

In view of (4.29), Lemma 4.3.3 can be invoked to conclude that for $k = 1, 2, \ldots, n$ there exists f_k such that with $f = f_k$ the integrand on the right is equal to $e_k = (0, \ldots 0, 1, 0 \ldots 0)'$. Then

$$\int_0^t f_k'(s)dv_s = \int_0^t e_k'dy_u = \int_0^t dy_u^k = y_t^k$$

This says that $y_t^k \in \mathcal{H}_t^v$. By the same argument $y_{t'}^k \in \mathcal{H}_{t'}^v \subset \mathcal{H}_t^v$ for any $t' < t$. Hence

$$\mathcal{L}\{y_s^k, k = 1, 2 \ldots n, 0 \leqslant s \leqslant t\} = \mathcal{H}_t^y \subset \mathcal{H}_t^v$$

This completes the proof.

4.3.5. Example. As a first example of the use of the innovations process in estimation problems, let us take

$n = 1$, $G(s) = 1$ and $z_t = \theta$, a single r.v., orthogonal to $\mathcal{H}^w$ with $E\theta = \mu$, $\text{var}(\theta) = \sigma^2$. This is the same problem as Example 3.3.3, except that we are no longer assuming $\{w_t\}$ to be normal. The problem of estimating θ from $\{y_s, s \leqslant t\}$ is most easily solved as in that example, but the following argument, using the innovations process, can be generalized to other situations and is in fact the basis for the derivation of Kalman filtering formulas in Section 4.4 below.

The observation equation is

$$dy_t = \theta\, dt + dw_t$$

and the corresponding innovations process is

$$d\nu_t = dy_t - \hat{\theta}_t dt$$

where $\hat{\theta}_t = \mathscr{P}_t^y(\theta - \mu) + \mu$. Now since $\mathcal{H}_t^\nu = \mathcal{H}_t^y$, Proposition 3.4.6 shows that

$$\hat{\theta}_t = \mu + \int_0^t \frac{d}{ds} E(\nu_s \theta)\, d\nu_s$$

Now let $p(t) = \|\theta - \hat{\theta}_t\|^2$. Then

$$E(\nu_s \theta) = E\left[\theta\left(\int_0^s (\theta - \hat{\theta}_u)\, du + w_s\right)\right]$$

$$= \int_0^s E(\theta - \hat{\theta}_u)^2\, du$$

$$= \int_0^s p(u)\, du$$

Thus

$$\hat{\theta}_t = \mu + \int_0^t p(s)\, d\nu_s \qquad (4.30)$$

We can use this to calculate $p(t)$. Indeed, from (4.30), using the properties of Wiener integrals,

$$E(\hat{\theta}_t - \mu)^2 = \int_0^t p^2(s)\, ds$$

Also

$$p(t) = E(\theta - \mu - \hat{\theta}_t + \mu)^2$$
$$= E(\theta - \mu)^2 + E(\hat{\theta}_t - \mu)^2 - 2E(\theta - \mu)(\hat{\theta}_t - \mu)$$
$$= E(\theta - \mu)^2 - E(\hat{\theta}_t - \mu)^2$$
$$= \sigma^2 - \int_0^t p^2(s)\,ds$$

Thus $p(t)$ satisfies the differential equation

$$\dot{p} = -p^2, p(0) = \sigma^2 \qquad (4.31)$$

to which, as is easily checked, the solution is

$$p(t) = \frac{1}{t + 1/\sigma^2} \qquad (4.32)$$

Now (4.30) can be written in the form

$$d\hat{\theta}_t = -p(t)\hat{\theta}_t dt + p(t)dy_t, \hat{\theta}_0 = \mu$$

which is a linear stochastic equation with the solution

$$\hat{\theta}_t = \Phi(t, 0)\mu + \int_0^t \Phi(t, s)p(s)dy_s \qquad (4.33)$$

where

$$\Phi(t, s) = \exp\left(-\int_s^t p(u)\,du\right)$$

But then

$$\frac{d}{ds}(\Phi(t, s)p(s)) = \Phi(t, s)(\dot{p} + p^2) = 0$$

so that

$$\Phi(t, s)p(s) = \Phi(t, t)p(t) = p(t) \qquad (4.34)$$

and

$$\Phi(t, 0)\mu = \Phi(t, 0)\frac{p(0)\mu}{\sigma^2} = \frac{p(t)\mu}{\sigma^2} \qquad (4.35)$$

Thus from (4.32) – (4.35)

$$\hat{\theta}_t = p(t)\left(\frac{\mu}{\sigma^2} + y_t\right)$$

$$= \frac{\frac{\mu}{\sigma^2} + y_t}{t + \frac{1}{\sigma^2}}$$

in agreement with (3.25). the salient feature is that (4.33) represents a recursive estimator for θ, since obviously for $t > \tau$

$$\hat{\theta}_t = \Phi(t, \tau)\hat{\theta}_\tau + \int_\tau^t \Phi(t, s)p(s)dy_s$$

and $\Phi(t, s)$ can be computed in terms of the solution of the differential equation (4.31). In more complicated cases it will not be possible to solve these equations in closed form for $\hat{\theta}_t$, but (4.31) and (4.33) are in a form suitable for numerical integration.

This example also illustrates the fact that the innovations process is a concept of some subtlety and can lead to surprising results. Suppose for example that $\mu = 0$ and $\sigma^2 = 1$, so that $\hat{\theta}_t = y_t/(1 + t)$. Then since $y_t = t\theta + w_t$, we have obtained *inter alia* the following result: if w_t is a BM and $\theta \sim N(0, 1)$, independent of $\{w_t\}$, then

$$v_t = w_t + \theta t - \int_0^t \frac{\theta s + w_s}{1 + s} \, ds$$

is also a BM. This does not seem immediately obvious!

Extension of the innovations results to the 'feedback' case

The purpose of this section is to show that the preceding results continue to hold if condition (C') is relaxed to (C), i.e. z_t and w_s may be correlated for $s \leq t$. The stronger condition (C') was only required in the proof of Lemma 4.3.2. Modification of this requires further examination of the result of Lemma 4.3.3. Denote by $L_2^n = L_2^n[a, b]$ the Hilbert space of functions $f: [a, b] \rightarrow R^n$ such that each component f^i is in L_2, with inner product

$$(f, g)_n = \sum_{i=1}^n (f^i, g^i)_{L_2}$$

and norm

$$\|f\|_n = \sqrt{(f, f)_n}.$$

Then Lemma 4.3.3 says that for any $f \in L_2^n$ there exists a unique $\phi \in L_2^n$ such that (4.25) holds, i.e. such that

$$f(t) = \phi(t) - \int_0^t h(t, s)\phi(s)\,ds. \qquad (4.25')$$

Now let I, H be the linear functions from $L_2^n \to L_2^n$ defined by

$$(I\phi)(s) = \phi(s), \qquad (H\phi)(s) = -\int_0^t h(t, s)\phi(s)\,ds.$$

Then the right-hand side of $(4.25')$ is just $(I + H)\phi$ and the content of Lemma 4.3.3 is that the function $I + H$ is one-to-one, so that the solution to (4.25) for given f is $\phi = (I + H)^{-1}f$. It also follows from the proof of the Lemma that $(I + H)^{-1}$ is bounded, i.e. there is a constant K such that

$$\| (I + H)^{-1}f \|_n \leqslant K \| f \|_n.$$

Returning now to Lemma 4.3.2, our objective is to show that, with

$$dy_t = z_t\,dt + G(t)\,dw_t,$$

each element X of $\mathcal{H}_t^y$ can be represented as $X = \int_0^t \beta'(s)\,dy_s$ for some $\beta \in L_2^n[0, t]$. Fix $t > 0$, and for $s \in [0, t]$ let $\bar{z}_s$ be the projection of z_s onto $\mathcal{H}_s^w$. By Proposition 4.4.1(b) this is given by

$$\bar{z}_s = \int_0^s \tilde{\gamma}(s, u)\,dw_u$$

for some $n \times r$ matrix-valued function $\tilde{\gamma}$. Now define

$$\gamma(s, u) = \tilde{\gamma}(s, u)G'(u)(G(u)G'(u))^{-1}.$$

Then $\gamma G = \tilde{\gamma}$, since G has rank n. Thus

$$\bar{z}_s = \int_0^s \gamma(s, u)G(u)\,dw_u$$

and

$$\text{cov}(\bar{z}_s) = \int_0^s \gamma(s, u)G(u)G'(u)\gamma'(s, u)\,du.$$

As before, measurable versions of $\{\bar{z}_s\}$ and of $\gamma(\cdot, \cdot)$ can be chosen. Now $E(z_s^i)^2$ is bounded on $[0, t]$ since $\{z_t\}$ is q.m. continuous, and $\text{var}(\bar{z}_s^i) \leqslant E(z_s^i)^2$. Using also the fact that

$G(u)G'(u) > bI_n$ we see that

$$\sum_{i,j} \int_0^t \int_0^s (\gamma^{ij}(s, u))^2 \, du \, ds < \infty.$$

Now define

$$z_s^* = z_s - \bar{z}_s.$$

Then $z_s^* \perp w_u$ for *all* s, u: for $u \leq s$ this is because $w_u \in \mathcal{H}_s^w$ and $z_s^* \perp \mathcal{H}_s^w$, whereas for $u > s$

$$(z_s^*, w_u) = (z_s^*, w_s) + (z_s^*, w_u - w_s)$$

and the second term is 0 from condition (C).

An arbitrary element X of $\mathcal{H}_t^y$ can be approximated as before by

$$X_m = \int_0^t \beta_m' \, dy_s = \int_0^t \beta_m' z_s^* \, ds + \int_0^t \beta_m' \bar{z}_s \, ds + \int_0^t \beta_m' G(s) \, dw_s$$

with step function integrands β_m. Denoting $\beta_{mp} = \beta_m - \beta_p$ we then have, since $z_s^* \perp \mathcal{H}_t^w$ for all s,

$$E(X_m - X_p)^2 = E\left[\int_0^t \beta_{mp}' z_s^* \, ds \right]^2$$

$$+ E\left[\int_0^t \beta_{mp}' \bar{z}_s \, ds + \int_0^t \beta_{mp}' G(s) \, dw_s \right]^2$$

$$\geq E\left[\int_0^t \beta_{mp}' \bar{z}_s \, ds + \int_0^t \beta_{mp}' G(s) \, dw_s \right]^2.$$

Now

$$\int_0^t \beta_{mp}' \bar{z}_s \, ds = \int_0^t \beta_{mp}'(s) \int_0^s \gamma(s, u) G(u) \, dw_u \, ds$$

$$= \int_0^t \int_u^t \beta_{mp}'(s) \gamma(s, u) \, ds \, G(u) \, dw_u$$

$$= \int_0^t \Gamma \beta_{mp}'(u) G(u) \, dw_u, \text{ say.}$$

Thus

$$E\left[\int_0^t \beta_{mp}' \bar{z}_s \, ds + \int_0^t \beta_{mp}' G(u) \, dw_u \right]^2$$

$$= E\left[\int_0^t (I + \Gamma)\beta'_{mp}(u)G(u)\mathrm{d}w_u\right]^2$$

$$= \int_0^t (I + \Gamma)\beta'_{mp}\, G(u)G'(u)(I + \Gamma)\beta_{mp}\mathrm{d}u$$

$$\geqslant b\,\|(I + \Gamma)\beta_{mp}\|_n^2$$

(using again the fact that $GG' \geqslant bI_n$). Let $\eta_{mp} = (I + \Gamma)\beta_{mp}$. We have shown that

$$\|\eta_{mp}\|_n^2 \;\leqslant\; \frac{1}{b}\,E(X_m - X_p)^2$$

so that $\|\eta_{mp}\| \to 0$ as $m, p \to \infty$ if $\{X_m\}$ is a Cauchy sequence. But according to our reformulation of Lemma 4.3.3

and
$$\beta_{mp} \;=\; (I + \Gamma)^{-1}\eta_{mp}$$

$$\|\beta_{mp}\|_n \;\leqslant\; K\,\|\eta_{mp}\|_n$$

for some constant K, so that $\{\beta_m\}$ is Cauchy in L_2^n and we conclude as before that

$$X \;=\; \lim_m X_m \;=\; \int_0^t \beta'(s)\mathrm{d}y_s$$

where $\beta = \lim_m \beta_m$. Thus Lemma 4.3.2, and hence Theorem 4.3.4, hold under conditions (A)–(D).

4.4. The Kalman Filter

The Kalman filter is a recursive scheme for estimating the state x_t of a dynamical system represented by the linear stochastic equation

$$\mathrm{d}x_t \;=\; A(t)x_t\mathrm{d}t + C(t)\mathrm{d}v_t, x_0 \;=\; x \qquad (4.36)$$

given an observed process y_t of the form

$$\mathrm{d}y_t \;=\; H(t)x_t\mathrm{d}t + G(t)\mathrm{d}w_t, y_0 \;=\; 0. \qquad (4.37)$$

$\{v_t\}$ and $\{w_t\}$ are o.i. processes spanning orthogonal subspaces and the initial r.v. x is orthogonal to $\{v_t, w_t\}$ If these are normal, then again (4.37) represents the 'signal in white noise' model

$$\dot{y}_t = H(t)x_t + G(t)\xi_t$$

The fact that a computationally tractable estimation scheme is available for this problem has had an enormous impact in a variety of applications, particularly in aerospace engineering. (Other major areas include, for example, econometrics, but there the discrete-time version is almost invariably used.) The result was first published [34, 35] in 1960 and 1961: excellent timing, since it coincided with the beginning of the space boom which as it happens provides the ideal field of application. The filtering procedure requires the coefficient matrices A, C, H, G to be known, and in many aerospace applications they can be calculated from well-specified system dynamics. In addition the noise, in, say, telemetry systems, tends to be white so that no difficult stochastic modelling problem in encountered there either.

The standing assumptions relative to (4.36), (4.37) are that $\{v_t\}$, $\{w_t\}$ have stationary orthogonal increments, $\mathcal{H}^v \perp \mathcal{H}^w$, the initial r.v. x is orthogonal to $\mathcal{H}^v$ and $\mathcal{H}^w$, and the elements of the matrices A, C, H, G are piecewise continuous time functions. The dimensions of x_t, y_t, v_t, w_t are n, m, l, r respectively. In addition $G(t)G'(t)$ is strictly positive definite for all t.

Let $\mathscr{P}_t$ denote the projection onto $\mathcal{H}_t^y$. For $\delta > 0$, $\mathscr{P}_t x_{t+\delta}$, $\mathscr{P}_t x_t$, $\mathscr{P}_t x_{t-\delta}$ are called the *predicted, filtered,* and *smoothed* linear estimates, respectively, given $\{y_s, s \leqslant t\}$. At first sight it seems that calculating these involves three separate problems, but it turns out that prediction and smoothing can be carried out in terms of filtered estimates, so the basic problem is to calculate $\hat{x}_t = \mathscr{P}_t x_t$. The filtering error is $\tilde{x}_t = x_t - \hat{x}_t$ and the innovations process corresponding to (4.37) is

$$d\nu_t = dy_t - H\hat{x}_t dt$$

$$= H\tilde{x}_t dt + G dw_t \qquad (4.38)$$

The conditions of the previous section are satisfied since $\mathcal{H}_t^x \perp \mathcal{H}_t^w$, so that $\{v_t\}$ is an o.i. process and

$$\text{cov}(\nu_t) = \int_0^t GG' ds.$$

Here now is the main result.

4.4.1. Theorem (Kalman filter). *Under the above assumptions, the filtered estimate $\hat{x}_t$ satisfies the linear stochastic equation*

$$d\hat{x}_t = (A - PH'(GG')^{-1}H)\hat{x}_t dt + PH'(GG')^{-1}dy_t$$

$$\hat{x}_0 = Ex = m_0. \tag{4.39}$$

Here $P(t) = E(\tilde{x}_t \tilde{x}_t')$ is the error covariance matrix; it satisfies the matrix Riccati equation

$$\dot{P} = CC' - PH'(GG')^{-1}HP + AP + PA' \tag{4.40}$$

$$P(0) = \mathrm{cov}(x) = P_0.$$

Remarks

The structure of (4.39) is brought out more clearly if it is written in the alternative form

$$d\hat{x}_t = A\hat{x}_t dt + PH'(GG')^{-1}d\nu_t$$

This shows that the filter has the same dynamics $A(t)$ as the state equation (4.36), and, like (4.36), is driven by an o.i. process, in this case $PH'(GG')^{-1}d\nu_t$ instead of the original $Cd\nu_t$.

Proof. The proof is in several stages. Assume to start with that $m(0) = 0$; then all r.v.'s have zero expectation.
1. It is convenient to define

$$D(t) = (GG')^{-1/2}$$

(that is, $D(t)$ is any matrix such that $DD' = D'D = (GG')^{-1}$) and

$$\pi_t = \int_0^t D(s)d\nu_s.$$

Then from Proposition 4.1.1,

$$\mathrm{cov}(\pi_t) = \int_0^t DGG'D'ds = I_m t$$

so that π_t has stationary o.i. Clearly $\mathcal{H}_t^\pi = \mathcal{H}_t^\nu$. Then since $\mathcal{H}_t^\nu = \mathcal{H}_t^y$ and $\hat{x}_t \in \mathcal{H}_t^y$, we have from Proposition 3.4.6,

$$\hat{x}_t = \int_0^t K(t,s)\,d\pi_s \qquad (4.41)$$

where

$$K(t,s) = \frac{d}{ds}E(x_t\pi_s')$$

2. Define

$$q_t = \hat{x}_t - \hat{x}_0 - \int_0^t A\hat{x}_s\,ds \qquad (4.42)$$

Then for $s < t$, $(q_t - q_s) \perp \mathcal{H}_s^y$.

This says in particular that $\{q_t\}$ is an o.i. process. This is true because the projection of dq_t onto $\mathcal{H}_t^y$ is the same as that of Cdv_t; but $Cdv_t \perp \mathcal{H}_t^y$ by assumption. Indeed

$$q_t - q_s = \hat{x}_t - \hat{x}_s - \int_s^t A\hat{x}_u\,du$$

and we have

$$\mathcal{P}_s\hat{x}_t = \mathcal{P}_s\,\mathcal{P}_t x_t = \mathcal{P}_s x_t$$

and

$$\mathcal{P}_s\int_s^t A\hat{x}_u\,du = \int_s^t A\,\mathcal{P}_s\,\mathcal{P}_u x_u\,du = \mathcal{P}_s\int_s^t A x_u\,du$$

Thus

$$\mathcal{P}_s(q_t - q_s) = \mathcal{P}_s\left(x_t - x_s - \int_s^t A x_u\,du\right)$$

$$= \mathcal{P}_s\left(\int_s^t Cdv\right) = 0$$

Since $\{q_t\}$ has o.i. with respect to $\{\mathcal{H}_t^y\}$ we can apply Theorem 3.4.5 and Proposition 4.1.1 (b) to conclude that there is a matrix-valued function g such that, for each t,

$$\sum_{i,j}\int_0^t (g^{ij}(s))^2\,ds < \infty$$

and

$$q_t = \int_0^t g(s)\,d\pi_s.$$

3. The next stage is to identify the function g. (4.42) can be rewritten as

$$d\hat{x}_t = A\hat{x}_t\,dt + g(t)\,d\pi_t$$

$$\hat{x}_0 = Ex_0 = 0 \qquad (4.43)$$

This equation has the solution

$$\hat{x}_t = \int\limits_0^t \Phi(t, s)g(s)\,\mathrm{d}\pi_s$$

where $\Phi(t, s)$ is the transition matrix of A. Comparing this with (4.41) we see that

$$\Phi(t, s)g(s) = K(t, s)$$

and evaluating this at $t = s$ gives

$$g(t) = \Phi(t, t)g(t) = K(t, t)$$

It remains to calculate this and to derive the equation for P.

4. From (4.38),

$$\pi_s = \int\limits_0^s DH\tilde{x}_u\,\mathrm{d}u + \int\limits_0^s DG\mathrm{d}w_u$$

Now $x_t \perp \mathcal{H}_s^w$, and hence

$$E(x_t\pi_s') = Ex_t \int\limits_0^s \tilde{x}_u' H'D'\mathrm{d}u$$

$$= \int\limits_0^s E(x_t\tilde{x}_u')H'D'\mathrm{d}u$$

For $t > u$,

$$x_t = \Phi(t, u)x_u + \int\limits_u^t \Phi(t, \tau)C\mathrm{d}v_\tau$$

The last term is orthogonal to $\mathcal{H}_u^{v,w}$ whereas $\tilde{x}_u \in \mathcal{H}_u^{v,w}$. Consequently

$$Ex_t\tilde{x}_u' = \Phi(t, u)Ex_u\tilde{x}_u'$$

$$= \Phi(t, u)E\tilde{x}_u\tilde{x}_u'$$

$$= \Phi(t, u)P(u). \qquad (4.44)$$

This shows that

$$Ex_t\pi_s' = \int\limits_0^s \Phi(t, u)P(u)H'(u)D'(u)\mathrm{d}u$$

so that

$$K(t, s) = \frac{d}{ds} E(x_t \pi'_s)$$

$$= \Phi(t, s)P(s)H'(s)D'(s)$$

and

$$g(t) = K(t, t) = PH'D'.$$

This completes the derivation of (4.39), apart from removing the assumption that $Ex_0 = 0$.

5. Write (4.43) in the form

$$d\hat{x}_t = A\hat{x}_t dt + PH'D'D(H\tilde{x}_t dt + Gdw_t)$$

Subtracting this from (4.36) gives

$$d\tilde{x}_t = (A - PH'D'DH)\tilde{x}_t dt + Cdv_t - PH'D'DGdw_t \tag{4.45}$$

Let $\Psi(t, s)$ be the transition matrix corresponding to $(A - PH'D'DH)$. Then the solution of this equation is

$$\tilde{x}_t = \Psi(t, 0)x_0 + \int_0^t \Psi(t, s)C(s)dv_s - \int_0^t \Psi(t, s)PH'D'DGdw_s$$

Now the three terms on the right are mutually orthogonal, so that

$$P(t) = E\tilde{x}_t \tilde{x}'_t = \Psi(t, 0)P_0\Psi'(t, 0) + \int_0^t \Psi(t, s)CC'\Psi'(t,s)ds$$

$$+ \int_0^t \Psi(t, s)PH'D'DHP\Psi'(t, s)ds$$

(4.40) is obtained by differentiating this with respect to t. Taking for example the second term,

$$\frac{d}{dt} \int_0^t \Psi CC'\Psi'ds = CC' + \int_0^t \left(\frac{\partial}{\partial t}\Psi\right)CC'\Psi'ds$$

$$+ \int_0^t \Psi CC'\left(\frac{\partial}{\partial t}\Psi'\right)ds$$

$$= CC' + (A - PH'D'DH)\int_0^t \Psi CC'\Psi'ds$$

$$+ \int_0^t \Psi CC'\Psi'ds(A' - H'D'DHP)$$

The other terms work similarly and we get

$$\dot{P} = CC' + PH'D'DHP + (A - PH'D'DH)P + P(A' - H'D'DHP)$$
$$= CC' - PH'D'DHP + AP + PA'$$

This is (4.40); the initial condition $P(0) = P_0$ is immediate from the expression for $P(t)$ above.

6. Finally, suppose $m(0) \neq 0$. Then $m(t) = Ex_t$ satisfies

$$\dot{m}(t) = Am(t), m(0) = m_0. \tag{4.46}$$

The centred process is $x_t^c = x_t - m(t)$ so that $\hat{x}_t^c$ satisfies (4.39) with $\hat{x}_0^c = 0$. Using (4.38) and (4.46) we find that $\hat{x}_t = \hat{x}_t^c + m(t)$ satisfies the same equation with $\hat{x}_0 = m(0) = m_0$. This completes the proof.

The reason why Theorem 4.4.1 gives a computationally effective solution to the filtering problem is that the covariance equation (4.40) is non-random. The function $P(t)$ can be calculated in advance and one then obtains the estimate $\{\hat{x}_t\}$ as the solution of a linear time varying equation driven by the observed process $\{y_t\}$. However, before carrying out this procedure we have to be assured that (4.40) cannot have more than one solution. (4.40) has a quadratic term and it is possible for such equations to 'blow up': for example, the solution to

$$\dot{p} = p^2, p(0) = 1$$

is

$$p(t) = \frac{1}{1 - t}$$

for $t < 1$, but there is no bounded solution on $[0, 2]$. However, this kind of thing cannot happen with (4.40) as long as the initial condition is non-negative definite, which, as $P(0) = \text{cov}(x)$, it always is.

4.4.2. Proposition. *The matrix Riccati equation*

$$\dot{Q} = CC' - QH'(GG')^{-1}HQ + AQ + QA'$$
$$Q(0) = P_0 \tag{4.47}$$

has a unique solution, which coincides with P(t) of Theorem 4.4.1. This solution is non-negative definite and bounded on any interval [0, t].

Proof. $P(t)$ is non-negative definite because $P(t) = \text{cov}(\tilde{x}_t)$ and bounded on $[0, t]$ since

$$E(\tilde{x}_s^i)^2 \leqslant E(x_s^i)^2.$$

$E(x_s^i)^2$ is uniformly bounded on $[0, t]$ due to the q.m. continuity of $\{x_s\}$.

The standard existence theorem for first order differential equations (see [24]) shows that (4.47) has a unique solution $Q(s)$ on some maximum interval $[0, t_0)$, $t_0 > 0$. Since $P(s)$ is a solution on that interval, $P(s) = Q(s)$ for $0 \leqslant s < t_0$. But then $Q(s) \rightarrow P(t_0)$ as $s \uparrow t_0$, so that $Q(s)$ is actually bounded on $[0, t_0]$. We can now use the existence theorem again, starting at t_0, to establish that there is a unique bounded solution on $[0, t_0 + \delta)$ for some $\delta > 0$, which contradicts the assumption that $[0, t_0)$ is the biggest such interval. Hence $Q(s)$ is unique and coincides with $P(s)$ on any interval $[0, t]$.

To clarify the content of Proposition 4.4.2, recall that in Example 4.3.5 the variance equation (4.31) was

$$\dot{p} = -p^2, \quad p(0) = \sigma^2$$

which has the bounded solution (4.32). Notice that the boundedness depends on the fact that the initial condition is positive: for a negative initial condition $p(0) = -\sigma^2$ the solution is $(t - 1/\sigma^2)^{-1}$ which is not bounded.

A slightly more complicated example is the following [12, Section 12.5].

4.4.3. Example. (Noisy observation of an Ornstein–Uhlenbeck process).

Here we have the following scalar equations:

$$dx = -\alpha x dt + dv$$

$$dy = x dt + g dw.$$

Suppose $Ex_0 = 0$, $\text{var}(x_0) = p_0$. Then the corresponding Kalman filter is

$$d\hat{x}_t = -\alpha \hat{x}_t dt + \frac{p(t)}{g^2}(dy - \hat{x}_t dt) \quad \hat{x}_0 = 0 \quad (4.48)$$

with the variance equation

$$\dot{p} = 1 - \frac{p^2}{g^2} - 2p\alpha, \quad p(0) = p_0 \qquad (4.49)$$

This can be solved explicitly, and we get

$$p(t) = p_1 + \frac{(p_1 - p_2)(p_0 - p_1)}{(p_0 - p_2)e^{2\beta t} - (p_0 - p_1)}$$

where

$$\beta = \sqrt{\alpha^2 + 1/g^2}, \quad p_1 = g^2(\beta - \alpha), \quad p_2 = -g^2(\beta + \alpha)$$

Note that $p(t) \to p_1$ as $t \to \infty$, so that, for large t, (4.48) is almost a time-invariant system $(p(t) \approx p_1)$. This happens even if $\{x_t\}$ is unstable, i.e. α is negative: in fact filtering is particularly easy in this case since eventually the 'signal' $x_t dt$ will completely swamp the 'noise' dw_t.

If $p(0) = p_1$ then $\dot{p} = 0$ (the other root of the right hand side of (4.49) is p_2, which is negative) so that $p(t) = p_1$ for all time. There are two opposing trends at work: on the one hand the observer learns more about $\{x_t\}$ as the observations continue for a longer time, and on the other the position of x_t gets less certain as it moves away from its initial position. The condition $p(0) = p_1$ ensures that these trends cancel out, leaving an exactly constant degree of uncertainty about $\{x_t\}$, reflected in constant error covariance $p(t) = p_1$. Notice that this is *not* the same thing as saying that $\{x_t\}$ is a stationary process (i.e. has distributions which do not change with time). The condition for this was given in Section 4.2 and is: $p(0) = 1/2\alpha$.

Prediction

This is straightforward. From (4.36) we have for $\delta > 0$

$$x_{t+\delta} = \Phi(t + \delta, t)x_t + \int\limits_t^{t+\delta} \Phi(t + \delta, s)C(s)\mathrm{d}v_s$$

Now the second term is orthogonal to $\mathcal{H}_t^y$, so that

$$\mathscr{P}_t^y x_{t+\delta} = \Phi(t + \delta, t)\hat{x}_t \qquad (4.50)$$

Let

$$\hat{x}_{t+\delta|t} = \mathscr{P}^y_t x_{t+\delta}$$

If A is time-invariant then (4.50) takes a particularly simple form because then

$$\Phi(t + \delta, t) = \Phi(\delta, 0).$$

Since $\Delta = \Phi(\delta, 0)$ is non-singular, $\hat{x}_{t+\delta|t}$ can be generated recursively as follows:

$$
\begin{aligned}
\mathrm{d}\hat{x}_{t+\delta|t} &= \mathrm{d}(\Delta\hat{x}_t) \\
&= \Delta\,\mathrm{d}\hat{x}_t \\
&= \Delta A\Delta^{-1}\hat{x}_{t+\delta|t} + \Delta PH'(GG')^{-1}\mathrm{d}\nu_t \\
\hat{x}_{\delta|0} &= \Delta m_0
\end{aligned}
$$

This just amounts to a change of coordinate system in the state space for x_t.

Smoothing

This is sometimes, and more descriptively, called *interpolation* and involves calculating

$$\hat{x}_{s|t} = \mathscr{P}^y_t x_s$$

for $s < t$. It is of course a non-causal operation in the sense that future values $\{y_u, s < u \leqslant t\}$ of the observations, as well as past values, are involved in estimating x_s. A wide variety of formulae for recursive and non-recursive smoothing is available in the literature for the three basic cases of fixed-point, fixed-lag and fixed-interval smoothing, which mean calculating $\hat{x}_{0|t}, \hat{x}_{t-\delta|t}$ and $\hat{x}_{t|b}$ respectively, where t is regarded as varying and δ, b as fixed. The reader is referred to [15, 33] for these, which are comparatively easy to work out, given the main result below which establishes the connection between filtering and smoothing by showing that the filtered estimate $\{\hat{x}_s, 0 \leqslant s \leqslant b\}$ contains all the information necessary to calculate any interpolation $\hat{x}_{t|b}$.

4.4.4. Theorem. *With standing assumptions and notation as before, the interpolation $\hat{x}_{t|b}(t < b)$ is differentiable in t and satisfies the differential equation*

$$\frac{d}{dt}\hat{x}_{t|b} = A(t)\hat{x}_{t|b} + C(t)C'(t)P^{-1}(t)(\hat{x}_{t|b} - \hat{x}_t)$$

with the terminal condition

$$\hat{x}_{b|b} = \hat{x}_b$$

Proof. (4.41) holds for all t, s and thus we can write

$$\hat{x}_{t|b} = \int_0^b K(t, s)\mathrm{d}\pi_s$$

$$= \int_0^t K(t, s)\mathrm{d}\pi_s + \int_t^b K(t, s)\mathrm{d}\pi_s$$

$$= \hat{x}_t + \int_t^b K(t, s)\mathrm{d}\pi_s$$

As before,

$$K(t, u) = E(x_t\tilde{x}_u')H'(u)D'(u)$$

but (4.44) is not the correct expression for $Ex_t\tilde{x}_u'$ if $u > t$. In this case we have from (4.45)

$$\tilde{x}_u = \Psi(u, t)\tilde{x}_t + z$$

Where z is a term orthogonal to $\mathcal{H}_t^v$. Consequently

$$Ex_t\tilde{x}_u' = E(x_t\tilde{x}_t')\Psi'(u, t)$$

$$= P(t)\Psi'(u, t)$$

and hence

$$\hat{x}_{t|b} = \hat{x}_t + P(t)\int_t^b \Psi'(s, t)H'(s)D'(s)\mathrm{d}\pi_s \qquad (4.51)$$

Let

$$\lambda_t = \int_t^b \Psi'(s, t)H'D'\mathrm{d}\pi_s$$

Now Ψ is the transition matrix of $\tilde{A} = (A - PH'D'DH)$. Using Proposition 4.2.1, we see that

$$\frac{\partial}{\partial t}\Psi'(s, t) = -\tilde{A}'(t)\Psi'(s, t)$$

so that $\tilde{\Psi}(t, s) = \Psi'(s, t)$ is the transition matrix corresponding to $-\tilde{A}'$. It follows that $\{\lambda_t\}$ is the solution of the stochastic equation

$$d\lambda_t = -\tilde{A}'(t)\lambda_t dt - H'(t)D'(t)d\pi_t \qquad (4.52)$$

$$\lambda(b) = 0$$

Now from Problem 4.5.2

$$d(P\lambda) = Pd\lambda + \dot{P}\lambda dt \qquad (4.53)$$

and combining (4.39), (4.40) and (4.51)–(4.53) gives

$$
\begin{aligned}
d\hat{x}_{t|b} &= d\hat{x}_t + Pd\lambda + \dot{P}\lambda dt \\
&= [A\hat{x}_t dt + PH'D'd\pi_t] - [P(A' - H'D'DHP)\lambda_t dt \\
&\quad + PH'D'd\pi_t] + [CC' - PHD'DH'P + AP + PA']\lambda_t dt \\
&= [A(\hat{x}_t + P\lambda_t) + CC'\lambda_t] dt \\
&= [A\hat{x}_{t|b} + CC'P^{-1}(\hat{x}_{t|b} - \hat{x}_t)] dt
\end{aligned}
$$

This gives the result.

It is perhaps worth pointing out that estimation of the state at varying times but with fixed data length gives differentiable sample functions but this is certainly not true for varying data lengths. We have just seen $(d/dt)\,\hat{x}_{t|b}$ exists for $t < b$; for $t > b$ we have of course

$$\hat{x}_{t|b} = \Phi(t, b)\hat{x}_b$$

so that

$$\frac{d}{dt}\hat{x}_{t|b} = A(t)\hat{x}_{t|b}$$

On the other hand (4.39) can be written in the form

$$\hat{x}_t = \hat{x}_0 + \int_0^t A\hat{x}_s ds + \int_0^t PH'D'DH\tilde{x}_s ds + \int_0^t PH'D'DGdw_s$$

which shows that $\hat{x}_t$ contains an additive o.i. process and hence cannot have differentiable sample functions.

Point process filtering

The application of Kalman filtering to point processes is based on the following observation. Suppose, for example, $\{X_t\}$ is a renewal process as in Proposition 3.2.2 and that the d.f. $G(\cdot)$ has density function $f(\cdot)$. Then the compensator $\{A_t\}$ of (3.12) is

$$A_t = \int\limits_0^t Z_s\,\mathrm{d}s$$

where

$$Z_t = \frac{f(t - T_{n-1})}{F(t - T_{n-1})}, \qquad t \in (T_{n-1}, T_n].$$

Since $Q_t = X_t - A_t$ is an o.i. process, we can write X_t in the form

$$\mathrm{d}X_t = Z_t\,\mathrm{d}t + \mathrm{d}Q_t \tag{4.54}$$

which puts $\{X_t\}$ in the 'signal plus noise' form required for the Kalman filter (but note that $\{Z_t\}$ and $\{Q_t\}$ are certainly not orthogonal).

In some cases the decomposition takes a form such that the filtering formulae (4.39) and (4.40) are directly applicable. Consider for example the problem of filtering a random telegraph signal $\{S_t\}$ defined by

$$S_t = \begin{cases} 0, & X_t \text{ even} \\ 1, & X_t \text{ odd}. \end{cases}$$

The compensating process $\{\bar{A}_t\}$ for $\{S_t\}$ is given in Problem 3.5.4. If $\{X_t\}$ is a Poisson process with rate λ then

$$\frac{\mathrm{d}}{\mathrm{d}t}\bar{A}_t = \begin{cases} \lambda, & X_t \text{ even} \\ -\lambda, & X_t \text{ odd} \end{cases}$$

$$= \lambda(1 - S_t) - \lambda S_t$$

$$= \lambda(1 - 2S_t).$$

Thus

$$\mathrm{d}S_t = -2\lambda S_t\,\mathrm{d}t + \lambda\,\mathrm{d}t + \mathrm{d}\bar{Q}_t \tag{4.55}$$

where $\{\bar{Q}_t\}$ is an o.i. process with variance λt (see Problem 3.5.4). This is in the form of a linear stochastic equation (4.36). As presently defined, $S_0 = 0$, but we could equally well consider random initial conditions $S_0 = 0, 1$ with probability $q, (1-q)$ (independent of $\{X_t\}$). Then from Theorem 4.2.4 $q(t) = \text{var}(S_t)$ satisfies

$$\dot{q} = -4\lambda q + \lambda \qquad q_0 = q(1-q).$$

Thus, as one might expect, $\lim_{t \to \infty} q(t) = \tfrac{1}{4}$ and $q(t) \equiv \tfrac{1}{4}$ if $q = \tfrac{1}{2}$.

Now suppose we observe $\{S_t\}$ in additive white noise, i.e. the observation process is

$$dY_t = S_t dt + \gamma dW_t$$

where γ is a constant and $\{W_t\}$ a standard BM. Applying the Kalman formulae (4.39) and (4.40) we find that the linear least squared estimator $\hat{S}_t$ is given by

$$d\hat{S}_t = \lambda(1 - 2\hat{S}_t)dt + p(t)(dY_t - \hat{S}_t dt)$$

$$\dot{p} = \lambda - \frac{1}{\gamma^2}p^2 - 4\lambda p \qquad p(0) = q(1-q).$$

The condition that (4.56) represents a time-invariant system is that

$$q(1-q) = p(0) = 2\lambda\gamma^2\left(\sqrt{1 + \frac{1}{4\lambda\gamma^2}} - 1\right).$$

(This is possible only if $p(0) \leqslant \tfrac{1}{4}$.)

It should be noted that the above argument depends on $\{X_t\}$ being a Poisson process since, for general renewal processes, (4.54) will not translate into a linear stochastic equation similar to (4.55) (but see Problem 4.5.10). Also in this particular case it is possible to calculate the best *non-linear* estimate $\pi_t = E[S_t | Y_s, s \leqslant t]$ directly from the non-linear filtering formula (6.9), using the fact that $S_t^2 = S_t$. We get

$$d\pi_t = \lambda(1 - 2\pi_t)dt + \pi_t(1 - \pi_t)(dY_t - \pi_t dt).$$

Since S_t is not a normal process, this is a better estimator than $\hat{S}_t$, i.e.

$$E(S_t - \pi_t)^2 < E(S_t - \hat{S}_t)^2 = p(t).$$

4.5. Problems and Complements

1. In Example 4.3.5 the inverse function generating y_t from $\{v_s, s \leqslant t\}$ is

$$y_t = \int_0^t \Phi(s, t) \, dv_s.$$

2. Let W_t be a process with stationary o.i., $g \in L_2[0, \infty)$ and $f(t) > 0$ be differentiable. Then the linear stochastic equation

$$dY_t = \frac{\dot{f}(t)}{f(t)} Y_t \, dt + f(t) g(t) \, dW_t, \qquad Y_0 = 0$$

has the solution

$$Y_t = f(t) \int_0^t g(s) \, dW_s.$$

This shows that if $X_t = \int_0^t g \, dW$ then

$$d(fX) = f \, dX + X \, df$$
$$= fg \, dW + X \dot{f} \, dt.$$

Now show that this formula also holds without the condition $f(t) > 0$.

3. Show that the pair (A, C) given by

$$A = \begin{bmatrix} 0 & 1 & & \\ & & 1 & \\ & & & \ddots & \\ & & & & 1 \\ a_1 & a_2 & \ldots & a_n \end{bmatrix} \qquad C = \begin{bmatrix} 0 \\ \vdots \\ 0 \\ 1 \end{bmatrix}$$

is controllable whatever the coefficients $(a_1 \ldots a_n)$. Interpret in terms of the behaviour of an nth order scalar differential equation with additive white noise.

4. Almost all time-invariant linear systems are controllable. There are several ways of interpreting this statement.
(a) An uncontrollable pair (A, C) can be made controllable by arbitrarily small perturbations of the elements of A and C, but not vice versa.

(b) A pair (A, C) can be regarded as a point in $R^{n(n+r)}$ (list the elements). The set of controllable systems is then open and dense in $R^{n(n+r)}$. (This is just another way of saying (a).)

(c) A *Bayesian* is someone who believes that the elements of (A, C) can be regarded as a random $n(n+r)$-vector whose distribution reflects his prior assessment of what they are likely to be. If this distribution is 'smooth' (has a density) then uncontrollable systems occur with probability 0.

According to the remarks in the proof of Proposition 4.2.2, it seems even less likely that a time-varying system will be uncontrollable. Formalize this.

5. If, in Example 4.4.3, $\{x_t\}$ is stationary, then 'uncertainty' is constant and 'learning' is taking place, so one expects that $p(t)$ will decrease with t. Show that indeed $1/2\alpha \geqslant p_1$, whatever the value of c.

6. Let $\{\beta_t\}$ be a standard n-vector BM and $S(t)$, $T(t)$ be $r \times n$ and $k \times n$ matrices whose elements are piecewise continuous functions of t. Let

$$\begin{bmatrix} v_t \\ w_t \end{bmatrix} = \int_0^t \begin{bmatrix} S(s) \\ T(s) \end{bmatrix} d\beta_s.$$

Assuming that $T(t)T'(t)$ is non-singular for all t, compute $\hat{v}_t = P_t^w v_t$ and $\tilde{v}_t = \hat{v}_t - v_t$, and show that $\{\tilde{v}_t\}$ is an o.i. process orthogonal to $\mathcal{H}^w$.

7. Let $P(t)$ be the solution of the matrix Riccati equation (4.40). Show that $P^{-1}(t)$ satisfies

$$\dot{P}^{-1} = -P^{-1}A - A'P^{-1} - P^{-1}CC'P^{-1} + H'(GG')^{-1}H.$$

Note that this is linear in P^{-1} if $C = 0$, i.e. if the state equation (4.36) is noise-free.

8. In the smoothing problem, let $\Sigma(t) = \text{cov}(\hat{x}_{t|b})$ (b is fixed). Show that

$$\Sigma(t) = P(t) - \int_t^b K(t, s)K'(t, s)ds$$

and hence (using Problem 7) that Σ satisfies

$$\dot{\Sigma} = (A + CC'P^{-1}) \Sigma + \Sigma (A + CC'P^{-1})' - CC'$$

$$\Sigma(b) = P(b).$$

9. For the smoothing problem, show that calculating $\hat{x}_s^B$, the projection of x_{b-s} onto $\mathcal{L}\{y_\tau, b - s \leqslant \tau \leqslant b\}$ can be formulated as a Kalman filter running backwards from time b. If $P_B(t)$ is the solution of the corresponding Riccati equation, show that

$$\hat{x}_{t|b} = \Sigma(t)(P^{-1}(t)\hat{x}_t + P_B^{-1}(b - t)\hat{x}_{b-t}^B)$$

and that

$$\Sigma^{-1}(t) = P^{-1}(t) + P_B^{-1}(b - t)$$

(see [33]).

10. Suppose T is a r.v. with density function $f(\cdot)$ and we observe a white noise process $\{\xi_t\}$ whose mean changes from 0 to m at time T, i.e. the observations process is

$$dY_t = m(1 - X_t)dt + dW_t$$

where $\{W_t\}$ is a BM and $X_t = I_{(t \geqslant T)}$. Using formula (3.11) show that the linear least-squares estimate of X_t can be obtained from a certain Kalman filter. (Note that T does not have to be exponential.)

CHAPTER FIVE

Linear stochastic control

For a system represented, as in Chapter 4, by a differential equation

$$\dot{x} = f(t, x_t, u_t, v_t),$$

an optimal control problem is specified by giving a *performance criterion*, which grades the possible control functions $\{u\}$ in order of preference by attaching a number $J(u)$ to each of them. $J(u)$ will be thought of as — and called — the *cost* of u, so that we want to choose the control which minimizes it. Of course if $J(u)$ actually represents profit then we maximize it by minimizing $-J(u)$, so the distinction between minimizing and maximizing is purely notational. In optimal control theory the type of cost function considered is almost invariably of the form

$$J(u) = E\left\{ \int_0^T h(t, x_t, u_t)\mathrm{d}t + g(x_T) \right\} \qquad (5.1)$$

where T is a not necessarily fixed, and possibly infinite, termination time. In engineering problems h is usually chosen to cost deviations from some desired trajectory of $\{x_t\}$ or the use of too much control force or energy, whereas g costs failure to reach some specified target set at the terminal time. In investment problems where $\{x_t\}$ represents the value of one's assets, h would be the consumption rate (one's income per unit time), and g is known, for obvious reasons, as the 'bequest function'.

It should be pointed out that, while it is clear that many deterministic (no disturbances) optimal control problems can reasonably be formulated so as to have a cost

$$\int_0^T h(t, x_t, u_t)\mathrm{d}t + g(x_T), \tag{5.2}$$

it is not at all obvious that an appropriate cost function for the stochastic case is obtained simply by averaging over all possible sample functions, as suggested by (5.1). For example, an investment strategy that resulted in bankruptcy with probability 0.8 would not be regarded by many people as satisfactory, even if it did maximize their *expected* wealth. However, often such objections can be taken care of by modifying the functions h, g so as to penalize undesirable events. Such questions are the province of *utility theory*, and it would be exceeding our brief to go into them here; see [25, 27].

An *open-loop* control is a function $u : [0, T] \to U$ where U is some (specified) allowable set of control values. A *closed-loop* or *feedback* control is one whose value at time t can also depend on the past evolution of the process $\{x_s\}$, i.e. is a function of the form $u(t) = u(t, \{x_s, 0 \leqslant s \leqslant t\})$. Specifying the exact form such dependence is allowed to take can be, as will be seen below, a problem in its own right, but we do not need to be more precise now. The fundamental difference between deterministic and stochastic control is that open-and closed-loop controls are equivalent in the deterministic but not in the stochastic case. The meaning of this is as follows. Clearly any open-loop control $u(\cdot)$ can be regarded as a closed-loop control of a degenerate kind. Conversely, suppose the system has deterministic dynamics $\dot{x} = f(t, x, u)$ and u is a feedback control such that the equation

$$\dot{x}_t = f(t, x_t, u(t, \{x_s, s \leqslant t\}))$$

has a unique solution. Now define $u^* : [0, T] \to U$ by

$$u^*(t) = u(t, \{x_s, s \leqslant t\})$$

and let $\{x_t^*\}$ be the solution of

$$\dot{x}_t^* = f(t, x_t^*, u_t^*)$$

Then obviously $x_t^* = x_t$ and consequently the cost (5.2) is the same for the open loop control u^* as for the closed-loop control u. It follows that the infimal cost using closed-loop controls is no smaller than that using open-loop controls, even though these are in principle a smaller class. However this statement is not true when random effects are present. The following example is completely artificial but does illustrate the characteristic features of stochastic control problems.

5.0.1. Example. This is a two-stage process, with initial state x_0 which is a r.v. taking the values 0, 1 with probability $\frac{1}{2}$ each, and with 'dynamics'

$$x_1 = x_0 + u_0.$$

The object is to minimize

$$J(u_0) = Ex_1^2.$$

In this case an open-loop control is a constant, $u_0 = \alpha$, and a closed-loop control is a function $u_0 = f(x_0)$. Now if $u_0 = -\frac{1}{2} + \alpha_0$ then

$$Ex_i^2 = \frac{1}{2}[(\frac{1}{2} + \alpha_0)^2 + (-\frac{1}{2} + \alpha_0)^2] = \frac{1}{4} + \alpha_0^2.$$

Thus the best open-loop control is $u_0 = -\frac{1}{2}$, with cost $\frac{1}{4}$, whereas obviously the best closed-loop control is $u_0 = -x_0$, with cost 0. So closed-loop controls are strictly better. However, there are also other possibilities. Suppose the controller can observe x_0, but not exactly. To be specific, suppose he observes

$$y = x_0 + w$$

where w has the values $-1, 0, +1$ with probability $\frac{1}{3}$ each, independent of x_0, and has to choose a function $u_0(y)$. The various possible outcomes are listed in Table 5.1.

Table 5.1

x_0	0	0	1	0	1	1
w	-1	0	-1	1	0	1
y	-1		0		1	2
Prob	$\frac{1}{6}$		$\frac{1}{3}$		$\frac{1}{3}$	$\frac{1}{6}$

We can now construct the optimal strategy to minimize $E(x_0 + u_0)^2$. If $y = -1, 2$ then x_0 must be 0, 1 respectively. If y is 0 or 1 then it is still equally likely that x_0 was 0 or 1. Thus the best strategy is

$$u_0(-1) = 0, \quad u_0(0) = u_0(1) = -\tfrac{1}{2}, \quad u_0(2) = -1.$$

It is easily checked that the corresponding cost is $\frac{1}{6}$. Using the same idea we could clearly define an observation y to yield an optimal cost anywhere between 0 and $\frac{1}{4}$.

This example illustrates the fact that in stochastic control the infimal cost is affected by the quality of the state measurements, closed- and open-loop controls being the extreme cases of zero and 'infinite' noise respectively. Notice that in the example the method of computing the optimal control was first to compute the *conditional distribution* of the state given the observations. This exemplifies the role of filtering in stochastic control.

There is basically only one stochastic control problem which lies within the ambit of the mean-square stochastic calculus of the preceding chapters, and this is the so-called *linear regulator problem*. The system dynamics are given by the stochastic equation

$$\mathrm{d}x_t = A(t)x_t\mathrm{d}t + B(t)u_t\mathrm{d}t + C(t)\mathrm{d}v_t \tag{5.3}$$

with observations

$$\mathrm{d}y_t = H(t)x_t\mathrm{d}t + G(t)\mathrm{d}w_t. \tag{5.4}$$

Here $\{v_t\}$ and $\{w_t\}$ will be taken to be independent vector Brownian motions,† so that (5.3), (5.4) have the

---

† The assumption of normality is not required except in connection with some remarks about the performance of linear control policies relative to possibly non-linear policies.

interpretation of the corresponding deterministic equations perturbed by additive white noise. $\{u_t\}$ takes values in R^k and the standing assumptions of Section 4.4 continue to hold. The control $\{u_t\}$ is to be chosen so as to minimize the cost

$$J(u) = E\left\{ \int_0^T (x_t'Q(t)x_t + u_t'R(t)u_t)dt + x_T'Fx_T \right\}. \quad (5.5)$$

This is a 'regulator' problem since evidently the objective of the controller is to drive the state to zero, i.e. make $|x_t|$ small. If u_t has the dimensions of, say, voltage, then $u_t'Ru_t$ is porportional to the instantaneous power, so the relative magnitudes of Q, F, R trade off good regulation on the one hand against low energy expenditure on the other. This problem formulation is particularly appropriate when $\{x_t\}$ actually represents the deviation from the nominal trajectory of some non-linear system and the overall objective is to keep this system as much as possible 'on target': any cost function $h(x_t, u_t)$ can be approximated by a quadratic as long as $|x_t|, |u_t|$ are small, so that (5.5) will always be a reasonable form of cost functions under this condition. If $|x_t|, |u_t|$ are large, then (5.3) and (5.4), if they are the linearized equations about a nominal trajectory, will in any case no longer be an adequate representation of the system dynamics.

There are several variants of the linear regulator problem. The general, 'partial observations' problem is to choose the control u_t as a function of the observations $\{y_s, s \leqslant t\}$ so as to minimize (5.5). The precise form such a function may take is considered in Section 5.3. The special case of 'complete observations' arises when the state can be measured in an error-free way, i.e. when $G(t) \equiv 0$ and $H(t) \equiv I_n$, so that $x_t = \dot{y}_t$. An even more special case is the deterministic one where $C(t) \equiv 0$ and the initial state x is fixed. Then the observations (5.4) are irrelevant and we are left with the problem of minimizing

$$\int_0^T (x_t'Qx_t + u_t'Ru_t)dt + x_T'Fx_T \qquad (5.6)$$

subject to the dynamics

$$\dot{x} = Ax + Bu. \qquad (5.7)$$

The main result is that the solutions to all three of these problems are in a sense the same. The optimal control for the deterministic problem (5.6), (5.7) is a linear feedback of the form $u_t = -K(t)x_t$, where the matrix-valued function $K(\cdot)$ can be computed in terms of the solution to a certain Riccati equation. This same control is also optimal for the completely observable stochastic problem, whereas in the partially observable case the optimal control is $u_t = -K(t)\hat{x}_t$, where K is as before and $\hat{x}_t$ is the minimum mean-square error estimate of x_t given $\{y_s, s \leqslant t\}$, which can be calculated using the Kalman filter. The conditional distribution of x_t will then be $N(\hat{x}_t, P(t))$, where $P(t)$ is as in (4.40). Thus u_t is being chosen in a sense as a function of the conditional distribution, which ties in with the ideas introduced in Example 5.0.1.

These neat results are entirely a consequence of the very special linear/quadratic structure of the problem. In fact, the 'control' and 'noise' aspects separate out in such a way that the completely observable regulator is hardly a stochastic problem at all. However, the stochastic aspects do of course play a significant role in the partially observable case.

As will be seen by reference to any standard text such as [14], the development of control theory has been in two fundamental streams, one based on the calculus of variations and leading to results such as Pontryagin's Maximum Principle, the other based on Bellman's Dynamic Programming. For various technical reasons, some of which are mentioned in Chapter 6, it is broadly true to say that the former has been more successful in deterministic control and the latter in stochastic control. As it happens, though, the linear regulator is one of the deterministic problems which is most readily solved by dynamic programming. We do this below and then introduce the stochastic elements in the subsequent sections.

5.1. Dynamic Programming and the Deterministic Linear Regulator

Consider a control system represented by the differential equation

$$\dot{x} = f(x, u) \qquad x(0) = x_0.$$

Note that time-varying functions $f(t, x, u)$ are included by adjoining $\dot{x}^{n+1} = 1, x^{n+1}(0) = 0$ so that $t = x^{n+1}(t)$. The cost function is

$$J(u) = \int_0^T h(x, u)dt + g(x_T).$$

We want to consider feedback controls $u : [0, T] \times R^n \to U \subset R^k$ where n, k are the dimensions of $x(t), u(t)$ respectively, and U is the permitted set of control values. Let $\mathcal{U}$ be the set of these functions such that the resulting system equation

$$\dot{x} = f(x, u(t, x))$$

has a unique solution on $[0, T]$. $\mathcal{U}$ is called the set of control *policies*, and it is not necessary at this juncture to define it more explicitly. For $u \in \mathcal{U}$ and $s, t \in [0, T]$, $u|_{[s, t]}$ is the restriction of u to the domain $[s, t] \times R^n$. $u^0 \in \mathcal{U}$ is *optimal* if

$$J(u^0) \leqslant J(u) \qquad \text{for all } u \in \mathcal{U}.$$

Suppose u^0 (with corresponding trajectory x_t^0) is optimal. Pick $t_1 \in [0, T]$ and let $x_1 = x_{t_1}^0$. Then u^0 restricted to $[t_1, T]$ must be optimal for the problem:

$$\text{minimize} \int_{t_1}^T h(x, u)dt + g(x_T)$$

$$\text{subject to } \dot{x} = f(x, u), x_{t_1} = x_1$$

because if some other control u' on $[t_1, T]$ achieved strictly lower cost, then the concatenation of u^0 on $[0, t_1]$ and u' on $[t_1, T]$ would give cost over $[0, T]$ strictly less than that achieved by u^0, which was however assumed to be optimal. This argument is known as Bellman's *principle of optimality*.

For any $(t, x) \in [0, T] \times R^n$ define

$$V(t, x) = \min_{u \mid [t, T]} \int_t^T h(x_s, u_s) ds + g(x_t)$$

$$\text{subject to } \dot{x} = f(x, u), x_t = x.$$

$V(t, x)$ is the *value function* for the control problem. (Of course, it has not been shown that the minimum actually exists.) Arguing heuristically, we can use the principle of optimality to derive an equation satisfied by $V(t, x)$. Let us *assume* that V exists and is differentiable in (t, x). We then have the Taylor expansion

$$V(s, y) = V(t, x) + \frac{\partial V}{\partial t}(t, x)(s - t)$$

$$+ \nabla_x V(t, x) \cdot (y - x) + o(|s - t|, |x - y|)$$

(5.8)

Now take any $\delta \in (0, T - t)$. Then

$$V(t, x) = \min_{u \mid [t, T]} \int_t^T h(x_s, u_s) ds + g(x_T)$$

$$= \min_{u \mid [t; T]} \left\{ \int_t^{t+\delta} h(x_s, u_s) ds + \int_{t+\delta}^T h(x_s, u_s) ds + g(x_T) \right\}$$

$$= \min_{u \mid [t, t+\delta)} \left\{ \int_t^{t+\delta} h(x_s, u_s) ds \right.$$

$$\left. + \min_{u \mid (t+\delta, T]} \int_{t+\delta}^T h(x_s, u_s) ds + g(x_T) \right\}$$

$$= \min_{u \mid [t, t+\delta)} \left\{ \int_t^{t+\delta} h(x_s, u_s) ds + V(t + \delta, x_{t+\delta}) \right\}$$

— invoking the principle of optimality.

Now using (5.8) we get

$$V(t, x) = \min_{u \in U} \left\{ h(t, x, u)\delta + V(t, x) + \frac{\partial V}{\partial t}(t, x)\delta \right.$$

$$\left. + \nabla_x V(t, x) f(x, y)\delta + o(\delta) \right\}.$$

Dividing by δ and letting $\delta \downarrow 0$ then gives

$$\frac{\partial V}{\partial t}(t, x) + \min_{u \in U} (\nabla_x V(t, x) f(x, u) + h(x, u)) = 0 \quad (5.9)$$

and

$$V(T, x) = g(x).$$

This is the so-called Bellman or Hamiltonian–Jacobi equation. It is a non-linear partial differential equation for $V(t, x)$ and the significance is that the original minimization over the set of control policies $\mathscr{U}$ has been replaced in (5.9) by a *pointwise* minimization over the space U of control values.

The above derivation depended on V being differentiable which we cannot know in advance. In order to prove something we *start* with (5.9) and work backwards. Here is the result.

5.1.1. Verification Theorem. *Suppose there exists a differentiable function V satisfying the Bellman equation and a control $u^0 \in \mathscr{U}$ which achieves the minimum in (5.9), i.e.*

$$\nabla_x V(t, x) f(x, u^0(t, x)) + h(x, u^0(t, x)) \leqslant \nabla_x V \cdot f(x, v)$$

$$+ h(x, v) \quad \text{for all } v \in U, (t, x) \in [0, T] \times R^n.$$

Then u^0 is optimal.

Proof. Let $u \in \mathscr{U}$ be any control and x_t the corresponding trajectory. Then from (5.9)

$$\frac{\partial V}{\partial t}(t, x) + \nabla_x V(t, x) \cdot f(x, u(t, x)) + h(t, x, u) \geqslant 0.$$

Thus

$$\frac{\mathrm{d}}{\mathrm{d}t}(V(t, x_t)) = \frac{\partial V}{\partial t}(t, x_t) + \nabla_x V(t, x_t) \cdot \dot{x}_t \geqslant -h(x_t, u_t).$$

Integrating, we have

$$V(s_2, x_{s_2}) - V(s_1, x_{s_1}) = \int_{s_1}^{s_2} \frac{\mathrm{d}}{\mathrm{d}t} V(t, x_t) \mathrm{d}t$$

$$\geqslant - \int_{s_1}^{s_2} h(x_t, u_t) \mathrm{d}t. \quad (5.10)$$

Now choose $s_2 = T, s_1 = 0$. Since $V(T, x) = g(x)$ we get

$$V(0, x_0) \leqslant \int_0^T h(x_s, u_s)\,ds + g(x_T) = J(u). \qquad (5.11)$$

With u^0, the same calculations apply, but with equality throughout in place of the inequality, to give

$$V(0, x_0) = J(u^0). \qquad (5.12)$$

But now (5.11) and (5.12) say that u^0 is optimal. This completes the proof.

If a function V satisfying the conditions of the verification theorem does exist then it is indeed the value function. This is shown by evaluating (5.10) at $s_2 = T$ and $(s_1, x(s_1)) = (t, x)$ (arbitrary). For this reason at most one function can satisfy these conditions, and it is unnecessary to stipulate that the solution of the Bellman equation be unique.

The verification theorem provides a *sufficient* condition for optimality. It is not a *necessary* condition and the main drawback of the dynamic programming approach is that in many problems the value function fails to be differentiable, so that the Bellman equation has no solution. In such cases the alternative approach, leading to necessary conditions for optimality via the Maximum Principle, yields more information.

Let us now apply the preceding results to the deterministic linear regulator problem (5.6)–(5.7). This is one case where the Bellman equation can be solved explicitly and the optimal policy determined. The control values are unconstrained so that $U = R^k$. Denoting $\nabla_x V = V_x$ and $V_t = \partial V/\partial t$ the Bellman equation is

$$\min_{u \in R^k} (x'Qx + u'Ru + V_t + V_x'(Ax + Bu)) = 0$$

i.e.
$$V(T, x) = x'Fx$$
$$x'Qx + V_t + V_x'Ax + \min_u (u'Ru + V_x'Bu) = 0.$$

Letting $V_x'B = C$ we have

$$u'Ru + Cu = (u + \tfrac{1}{2}R^{-1}C')'R(u + \tfrac{1}{2}R^{-1}C') - \tfrac{1}{4}CR^{-1}C'$$

Thus the minimum is $-\frac{1}{4}CR^{-1}C'$ and is achieved at

$$u^0 = -\frac{1}{2}R^{-1}C' = -\frac{1}{2}R^{-1}B'V_x$$

The Bellman equation becomes

$$V_t + x'Qx + V_x'Ax - \frac{1}{4}V_x'BR^{-1}B'V_x = 0 \qquad (5.13)$$

$$V(T, x) = x'Fx.$$

We can verify that there is a solution to this in the quadratic form

$$V(t, x) = x'\Lambda(t)x + \beta(t) \qquad (5.14)$$

($\Lambda(t)$ symmetric.) With V given by (5.14) we have

$$V_t(t, x) = x'\dot{\Lambda}x + \dot{\beta}$$

$$V_x(t, x) = 2\Lambda x$$

so that substituting in (5.13) gives

$$x'Qx + (x'\dot{\Lambda}x + \dot{\beta}(t)) + 2x'\Lambda Ax - x'\Lambda BR^{-1}B'\Lambda x = 0$$

i.e. $x'(\dot{\Lambda} + Q + 2\Lambda A - \Lambda BR^{-1}B'\Lambda)x + \dot{\beta}(t) = 0.$

Noting that $x'\Lambda Ax = x'A'\Lambda x$ we see that (5.14) is indeed a solution if $\beta(t) = 0$ and Λ satisfies the Riccati equation

$$-\dot{\Lambda} = Q + \Lambda A + A'\Lambda - \Lambda BR^{-1}B'\Lambda \qquad (5.15)$$

$$\Lambda(T) = F.$$

We want to show this has a bounded solution (as remarked earlier, no uniqueness is required). This is done by exploiting the fact that the same equation appears in connection with a certain Kalman filter, where a bounded solution is assured.

Indeed, recall from Theorem 4.4.1 that for the Kalman filter model:

$$dx_t = \tilde{A}x_t dt + Cdv_t$$

$$dy_t = Hx_t dt + Gdw_t,$$

the error covariance matrix $P(t)$ satisfies

$$\dot{P} = CC' + \tilde{A}P + P\tilde{A}' - PH(GG')^{-1}H'P$$

$$P(0) = \text{cov}(x_0)$$

and $P(t)$ is bounded by the unconditional covariance $E(x_t x_t')$.

Now choose the following coefficients:

$$\tilde{A} = A'$$
$$H = B'$$
$$C = \sqrt{Q} \qquad \text{cov}(x_0) = F \qquad (5.16)$$
$$G = \sqrt{R}$$

and reverse the time, i.e. let $\tau = T - t$. Then $\tilde{P}(\tau) = P(T - \tau)$ satisfies

$$\frac{d}{d\tau}\tilde{P}(\tau) = Q + A'\tilde{P} + \tilde{P}A - \tilde{P}BR^{-1}B'\tilde{P}$$

$$\tilde{P}(T) = F.$$

But this is precisely the Riccati equation (5.15) of the control problem. Hence this has a bounded solution $\Lambda(t)$. Applying the verification theorem we conclude that the optimal control is

$$u^0(t, x) = -\tfrac{1}{2}R^{-1}B'V_x$$
$$= -R^{-1}(t)B'(t)\Lambda(t)x$$

and the cost of this policy is

$$J(u_0) = V(0, x_0) = x_0'\Lambda(0)x_0.$$

A very important feature of this result is that the optimal policy is linear in x, though the admissible class $\mathscr{U}$ contains much more than just linear controls. In fact $u \in \mathscr{U}$ if the equation

$$\dot{x}(t) = Ax(t) + Bu(t, x(t)), x(0) = x_0$$

has a unique solution, so that certainly all functions which are Lipschitz continuous in x, i.e. such that

$$|u(g, x) - u(t, y)| \leqslant K|x - y|,$$

for some constant K, are included.

The preceding results show that to every linear regulator there corresponds a Kalman filter featuring the same Riccati equation, and conversely. The filter and regulator are related by (5.16) and are said to be *dual* to each other.

5.2. The Stochastic Linear Regulator

Quadratic Integral Expressions

In order to 'stochasticize' the linear regulator problem we need to consider expressions of the form

$$V(t, x) = E_{t,x} \left\{ \int_t^T x_s' M(s) x_s \, ds + x_T' F x_T \right\} \quad (5.17)$$

where $\{x_t\}$ is the solution of the linear stochastic equation

$$dx_s = A(s) x_s \, ds + C(s) dv_s \qquad s \in [t, T] \quad (5.18)$$

$$x_t = x \qquad \text{(fixed)}.$$

The subscript (t, x) in (5.17) refers to the initial condition in (5.18). Note first that if $Q(s) = Ex_s x_s'$, then

$$E(x_s' M(s) x_s) = \text{tr}(M(s)Q(s)) = \text{tr}(Q(s)M(s)) \quad (5.19)$$

because each of these expressions is equal to

$$\sum_{i,j} (Ex_s^i x_s^j) M^{ij}(s).$$

Now the solution to (5.18) is

$$x_s = \Phi(s, t) x + \int_t^s \Phi(s, u) C dv_u$$

where Φ is the transition matrix corresponding to A. Thus using (5.19) and the fact that $E \int (\cdot) dw = 0$ we have

$$Ex_s' Mx_s = x' \Phi'(s, t) M(s) \Phi(s, t) x$$
$$+ \text{tr} \left\{ M(s) \int_t^s \Phi(s, u) C(u) C'(u) \Phi'(s, u) du \right\}$$

and hence

$$V(t, x) = x' \left\{ \int_t^T \Phi'(s, t) M(s) \Phi(s, t) ds + \Phi'(T, t) F \Phi(T, t) \right\} x$$
$$+ \int_t^T \text{tr} \left\{ M(s) \int_t^s \Phi(s, u) C(u) C'(u) \Phi'(s, u) du \right\} ds$$
$$+ \text{tr} \left\{ F \int_t^T \Phi(T, s) CC' \Phi'(T, s) ds \right\} \quad (5.20)$$

Denote the first term by $x' N(t) x$. Now differentiating N

gives, using Proposition 4.2.1(d),

$$
\begin{aligned}
\dot{N} &= -M + \int_t^T \frac{\partial}{\partial t}\,\Phi'M\Phi\,ds + \frac{\partial}{\partial t}\,\Phi'(T, t)F\Phi(T, t) + \dots \\
&= -M - A'\left(\int_t^T \Phi'M\Phi\,ds + \Phi'(T, t)F\Phi(T, t)\right) + \dots \\
&= -M - A'N - NA.
\end{aligned}
$$

Thus $N(t)$ satisfies the equation:

$$
\dot{N} + A'N + NA + M = 0
$$
$$
N(T) = F.
$$

The second term in (5.20) is, after interchanging the order of integration:

$$
\int_t^T \int_u^T \operatorname{tr}\{(M(s)\Phi(s, u)C(u))(C'(u)\Phi'(s, u))\}\,ds\,du
$$

$$
= \int_t^T \int_u^T \operatorname{tr}\{C'(u)\Phi'(s, u)M(s)\Phi(s, u)C(u)\}\,ds\,du \ \text{(using (5.19))}
$$

$$
= \int_t^T \operatorname{tr}\left\{C'(u)\left\{\int_u^T \Phi'(s, u)M(s)\Phi(s, u)\,ds\right\}C(u)\right\}du.
$$

Similarly the third term is

$$
\int_t^T \operatorname{tr}\{C'(u)\Phi'(T, u)F\Phi(T, u)C(u)\}\,du
$$

so that denoting the sum of the second and third terms by $q(t)$, we have

$$
q(t) = \int_t^T \operatorname{tr}(C'(u)N(u)C(u))\,du.
$$

We have shown the following.

5.2.1. Proposition. *Let* $V(t, x) = E_{t, x}\left\{\int_t^T x_s'M(s)x_s\,ds + x_T'Fx_T\right\}$ *where* x_t *is given by (5.18) above. Then*

$$
V(t, x) = x'N(t)x + \int_t^T \operatorname{tr}(C'NC)\,ds \tag{5.21}
$$

where N *is the unique solution of*

$$
\dot{N} + A'N + NA + M = 0, \qquad N(T) = F.
$$

5.2.2. Corollary. *Suppose in (5.18) that the initial point x is a r.v. orthogonal to $\mathcal{H}^v$, with mean m and covariance Q. Then*

$$E\left\{\int_t^T x_s' M(s) x_s \, ds + x_T' F x_T\right\} = EV(t, x)$$

$$= m'N(t)m + \text{tr}\left\{QN(t) + \int_t^T C'NC \, ds\right\}$$

Proof. This follows the same argument as above, using the orthogonality of x and $\mathcal{H}^v$ in calculating $Ex_s'Mx_s$ and then formula (5.19) with $Exx' = Q + mm'$. ($V(t, x)$ is in fact the conditional expectation of the quadratic expression given x.)

These results can be rephrased in the following way. For any function $f: [0, T] \times R^n \to R$ which is once differentiable in t and twice in x, define the operator $\mathcal{L}$ by

$$(\mathcal{L}f)(t, x) = f_t + f_x' Ax + \tfrac{1}{2}\text{tr}\{C'(t)f_{xx}C(t)\} \quad (5.22)$$

where

$$f_t = \frac{\partial f}{\partial t}, \quad (f_x)^i = \frac{\partial f}{\partial x_i}, \quad (f_{xx})^{ij} = \frac{\partial^2 f}{\partial x_i \partial x_j}.$$

In the case of $V(t, x)$ we have from (5.21)

$$V_t = x'\dot{N}x - \text{tr}(C'NC), \quad V_x = 2Nx, \quad V_{xx} = 2N$$

and therefore

$$\mathcal{L}V(t, x) = x'(-A'N - NA - M)x - \text{tr}(C'NC)$$
$$+ 2x'NAx + \text{tr}(C'NC)$$
$$= -x'Mx.$$

Thus Proposition 5.2.1 is equivalent to saying that V satisfies the equation

$$V(t, x) = -E_{t,x}\left[\int_t^T \mathcal{L}V(s, x_s) \, ds - x_T' F x_T\right]. \quad (5.23)$$

This is a special case (restricted to quadratic functions) of the so-called 'Ito differential rule' of stochastic calculus, which gives a similar formula for any differentiable function V. But to interpret this a more general type of stochastic integral than the Wiener integral we have considered so far is required. Some further details will be found in Chapter 6.

Let us now turn to the stochastic linear regulator problem, i.e. the minimization of

$$J(u) = E\left[\int_0^T (x_t'Qx_t + u_t'Ru_t)dt + x_T'Fx_T\right]$$

where $\{x_t\}$ is the solution of the stochastic equation

$$dx_t = Ax_tdt + Bu_tdt + Cdv_t \qquad (5.24)$$

$$x_0 = x^0.$$

The initial r.v. x^0 is $N(m_0, P_0)$ distributed, independent of $\{v_t\}$.

5.2.3. Definition. Let $\mathcal{K}$ denote the set of $n \times m$ matrix valued functions on $[0, T]$ with piecewise continuous elements. Then $u(t, x_t)$ is an *admissible feedback control* if

$$u(t, x_t) = K(t)x_t$$

for some $K \in \mathcal{K}$. We also write $u \in \mathcal{K}$ and $J(u) = J(K)$ in this case.

The admissible controls have to be linear in x so that (5.24) is a linear stochastic equation for $u \in \mathcal{K}$. Since the evolution of (5.24) from t onwards only depends on x_t and $\{v_s, s \geqslant t\}$ is seems likely that there would be no advantage in using general past-dependent controls $u(t, \{x_s, s \leqslant t\})$ instead of $u(t, x_t)$, and indeed there is none; see Problem 5.4.1. We shall see in Section 5.3 how it is possible to incorporate such past dependence without leaving the framework of linear equations.

With $u \in \mathcal{K}$, the system equation (5.24) becomes

$$dx_t = \tilde{A}x_tdt + Cdw_t \qquad (5.25)$$

where

$$\tilde{A}(t) = A(t) + B(t)K(t)$$

so that the solution is well-defined for all $K \in \mathcal{K}$.

Let $K_0 \in \mathcal{K}$, let x_t^0 be the solution of (5.25) with $\tilde{A} = A + BK_0$ and let

$$V^0(t, x) = E_{t,x}\left(\int_t^T (x_s^{0'}Qx_s^0 + u_s^{0'}Ru_s^0)ds + x_T^{0'}Fx_T^0\right)$$

$$= E_{t,x}\left(\int_t^T x_s^{0'}M^0x_s^0ds + x_T^{0'}Fx_T^0\right)$$

where
$$M^0 = Q + K_0'RK_0.$$

Finally, for $K \in \mathcal{K}$, let $\mathcal{L}_K$ denote the operator defined by (5.22) but with A replaced by $A + BK$, and let

$$M_K = Q + K'RK$$

so that
$$x'Qx + u'Ru = x'M_K x.$$

5.2.4. Theorem. *Suppose $K_0 \in \mathcal{K}$ has the property that*

$$0 = \mathcal{L}_{K_0} V^0(t, x) + x'M_{K_0} x \leqslant \mathcal{L}_K V^0(t, x) + x'M_K x \quad (5.26)$$

for all $K \in \mathcal{K}$ and $(t, x) \in [0, T] \times R^n$.
Then K_0 is optimal in $\mathcal{K}$.

Proof. By definition
$$V^0(t, x) = E_{t,x}\left(\int_t^T x_s^{0\prime} M_{K_0} x_s^0 ds + x_T^{0\prime} F x_T^0\right). \quad (5.27)$$

Now let $K \in \mathcal{K}$ be arbitrary and let x_t denote the corresponding solution of (5.25). Then from (5.26)

$$0 = -\mathcal{L}_{K_0} V^0 - x'M_{K_0} x \geqslant -\mathcal{L}_K V^0 - x'M_K x, \quad x \in R^n.$$

Applying the formula (5.23) we have

$$V^0(t, x) = -E_{t,x}\left\{\int_t^T (\mathcal{L}_K V^0(s, x_s)ds - x_T' F x_T\right\}$$

$$\leqslant -E_{t,x}\left\{\int_t^T (\mathcal{L}_{K_0} V^0 + x_s' M_{K_0} x_s \right.$$

$$\left. - x_s' M_K x_s)ds - x_T' F x_T\right\}$$

$$= -E_{t,x}\left\{\int_t^T (-x_s' M_K x_s)ds - x_T' F x_T\right\}$$

$$= E_{t,x}\left\{\int_t^T (x_s' Q x_s + u_s' R u_s)ds + x_T' F x_T\right\}.$$

$$= V(t, x),$$

where V is similar to V^0 but with K replacing K_0. Taking $(t, x) = (0, x_0)$ and using (5.27) and Corollary 5.2.2 we see that

$$E V^0(0, x_0) = J(K_0) \leqslant J(K).$$

Thus K_0 is optimal in $\mathcal{K}$, as claimed.

Remarks

The preceding result is the stochastic version of Theorem 5.1.1. By definition

$$\mathcal{L}_K V = \tfrac{1}{2}\mathrm{tr}(C'V_{xx}C) + V_x(A + BK)x + V_t$$
$$= \tfrac{1}{2}\mathrm{tr}(C'V_{xx}C) + V_x(Ax + Bu) + V_t.$$

Then equation (5.26) above says

$$\min_K(V_t + \tfrac{1}{2}\mathrm{tr}(C'V_{xx}C) + V_x(A - BK)x + x'(Q + K'RK)x) = 0.$$
$$(5.28)$$

For comparison with the deterministic case let us substitute

$$u = Kx$$
$$f(x, u) = Ax + Bu$$
$$h(x, u) = x'Qx + u'Ru.$$

Then (5.28) becomes:

$$V_t + \tfrac{1}{2}\mathrm{tr}(C'V_{xx}C) + \min_u(V_x \cdot f + h) = 0.$$

This is the stochastic Bellman equation; but notice in the theorem we could only use it to establish optimality of a candidate u in case V is *quadratic* and u is *linear* $(= Kx)$.

Comparing with the Bellman equation (5.9) (valid for any differentiable function U and control u) we notice that (5.28) has an extra second-order term $\tfrac{1}{2}\mathrm{tr}(C'V_{xx}C)$.

Let us now show that there exist V^0, K^0 satisfying (5.26), which are consequently optimal. Write (5.28) in the form

$$V_t^0 + \tfrac{1}{2}\mathrm{tr}(C'V_{xx}^0C) + \min_K(V_x^{0'}(A + BK)x + x(Q + K'RK)x) = 0.$$

Putting $u = Kx$ this becomes

$$V^0 + \tfrac{1}{2}\mathrm{tr}(C'V_{xx}^0C) + x'Qx + V_x^0Ax + \min(u'Ru + V_x^{0'}Bu) = 0.$$

and

$$u'Ru + V_x^{0'}Bu = (u + \tfrac{1}{2}R^{-1}B'V_x^0)'R(u + \tfrac{1}{2}R^{-1}B'V_x^0)$$
$$- \tfrac{1}{4}V_x^{0'}BR^{-1}B'V_x^0.$$

Thus (if V^0 is quadratic) the minimum is achieved at

$$u = - \tfrac{1}{2} R^{-1} B' V_x^0$$

and finally the Bellman equation is

$$V_t^0 + \tfrac{1}{2} \mathrm{tr}(C' V_{xx}^0 C) + x'Qx + V_x^0{}' Ax - \tfrac{1}{4} V_x^0{}' B R^{-1} B' V_x^0 = 0$$

$$V(T, x) = x'Fx. \tag{5.29}$$

We now see whether this can be satisfied by a quadratic function

$$V^0(t, x) = x'S(t)x + s(t) \tag{5.30}$$

so that

$$V_t^0 = x'\dot{S}x + \dot{s}$$

$$V_x^0 = 2Sx$$

$$V_{xx}^0 = 2S.$$

Substituting in (5.29) gives

$$[\dot{s} + \mathrm{tr}(C'SC)] + x'\{\dot{S} + SA + A'S + Q - S'BR^{-1}B'S\}x = 0.$$

Thus (5.30) is a solution as long as S, s satisfy

$$\dot{S} + SA + A'S + Q - S'BR^{-1}B'S = 0 \tag{5.31}$$

$$S(T) = F$$

$$\dot{s} = - \mathrm{tr}(C'SC), \quad s(T) = 0. \tag{5.32}$$

But this is the same Riccati equation that appeared in the deterministic linear regulator problem (equation (5.15)) where it was shown to have a bounded solution $\Lambda(t)$. Summarizing:

5.2.5. Theorem. *Let $\Lambda(t)$ be the bounded symmetric solution of the Riccati equation (5.15). Then the control*

$$u^0(t) = -R^{-1}B'\Lambda x$$

is optimal for the stochastic linear regulator problem in the class $\mathcal{K}$. Its cost is

$$J(u_0) = \int_0^T \mathrm{tr}(C'(s)\Lambda(s)C(s))\,\mathrm{d}s + m_0'\Lambda(0)m_0 + \mathrm{tr}(P_0\Lambda(0)).$$

The same control is also optimal for the deterministic linear regulator.

Proof. Everything has been established except the expression for the optimal cost. Now

$$E[J(u^0)|x_0 = x] = V^0(0, x_0) = x'\Lambda(0)x + s(0).$$

This gives $J(u^0)$ on noting that

$$s(0) = \int_0^T \text{tr}(C'\Lambda C)\,ds$$

and

$$E(x_0'\Lambda(0)x_0) = m_0'\Lambda(0)m_0 + E((x_0 - m_0)'\Lambda(0)(x_0 - m_0))$$

$$= m_0'\Lambda(0)m_0 + \text{tr}(P_0\Lambda(0)).$$

Note that the 'noise' coefficient C does not affect Λ but simply contributes the term $s(0)$ to $J(u^0)$. Thus we recover the deterministic result simply by setting $C(t) = P_0 = 0$. But Theorem 5.1.1 is stronger because the class of controls within which u^0 was shown to be optimal is bigger: it includes some non-linear functions. The argument given above, which was borrowed from Wonham [20], is designed to be as close as possible to that used in non-linear problems. If we interpret non-linear stochastic equations in terms of Ito calculus (see Chapter 6) then it turns out that formula (5.23) is valid for all smooth functions, not just for quadratics, and we can go on to show that our candidate u^0 is optimal in a class of controls essentially similar to that considered in the deterministic case.

Theorem 5.2.5 is simultaneously one of the most satisfactory and one of the most disappointing results in stochastic control. It is satisfactory in the sense that the presence of noise only degrades performance but does not qualitatively change the optimal control function, i.e. the optimal policy is 'stable' under noise perturbations, a desirable engineering property. The disappointment is of course that we have developed a whole theory of stochastic systems only to find that the most simple-minded idea for dealing with noise, namely to apply the optimal deterministic control for the noise set at its mean value of zero, is actually optimal. But it must be remembered that this result depends entirely on the linear/quadratic assumptions. The truth of the matter is that the problem is far too special to bring out any specifically stochastic features; we have to wait for the partially observable case for that.

Open-Loop Control for the Stochastic Linear Regulator

In the preceding section the admissible controls $\mathscr{K}$ were of the form $u = Kx$, i.e. feedback controls. Now let us consider open-loop controls, i.e. (deterministic) functions $u : [0, T] \to R^k$. According to the argument given at the beginning of the chapter, it should not be possible to achieve lower cost with open-loop controls. Let us show this. The system is, as before,

$$dx_t = Ax_t dt + Bu(t)dt + Cdv_t$$

with x_0 independent of $\{v_t\}$ and $Ex_0 = m_0$, $\mathrm{cov}(x_0) = P_0$.

For an open-loop control u, split up x_t into its 'deterministic' part x_t^* and its 'stochastic' part $\bar{x}_t$ as follows:

$$dx_t^* = Ax_t^* dt + Bu(t)dt \qquad x_0^* = m_0 \qquad (5.33)$$

$$d\bar{x}_t = A\bar{x}_t dt + Cdv_t \qquad \bar{x}_0 = x_0 - m_0.$$

We then have $E\bar{x}_t = 0$ and $x_t = \bar{x}_t + x_t^*$, so that

$$Ex_t' Qx_t = E\bar{x}_t' Q\bar{x}_t + x_t^{*'} Qx_t^*.$$

Thus the cost for the linear regulator with deterministic control is

$$J(u) = \int_0^T (x_t^{*'} Qx_t^* + u_t' Ru_t')dt + x^* Fx^* + E \int_0^T \bar{x}_t' Q\bar{x}_t dt$$
$$+ E\bar{x}_T' F\bar{x}_T \qquad (5.34)$$

The last two terms do not depend on u, and from Proposition 5.2.1, their value is, for given x_0.

$$(x_0 - m_0)' N(0)(x_0 - m_0) + \int_0^T \mathrm{tr}(C'NC)dt$$

where N is the solution of

$$\dot{N} + A'N + NA + Q = 0 \qquad (5.35)$$

$$N(T) = F.$$

Thus taking into account the distribution of x_0, the contribution of the last two terms is

$$\int_0^T \text{tr}(C'NC)\,dt + \text{tr}(P_0 N(0)).$$

Now the control problem formed by dynamics (5.33) and cost (5.34) is just the deterministic regulator problem. The solution is

$$u^{0l}(t) = -R^{-1}B\Lambda x_t^* \qquad (5.36)$$

where Λ satisfies the Riccati equation (5.15):

$$\dot{\Lambda} + A'\Lambda + \Lambda A + Q - \Lambda'BR^{-1}B'\Lambda = 0 \qquad (5.37)$$
$$\Lambda(T) = \Lambda.$$

Notice that (5.36) is *not* a feedback control for the present problem but must be computed 'off-line'. With control (5.36) the value of the first two terms in (5.34) is $m_0'\Lambda(0)m_0$, so the total cost is

$$J(u^{0l}) = \int_0^T \text{tr}(C'NC)\,dt + \text{tr}(P_0 N(0)) + m_0'\Lambda(0)m_0.$$

It is the same as $J(u^0)$ except that $\Lambda(t)$ has been replaced by $N(t)$ in the first two terms. Let $M_t = N_t - \Lambda_t$ and $Y_t = \Lambda'BR^{-1}B'\Lambda$. Subtracting (5.37) from (5.35) shows that M satisfies

$$\dot{M} + A'M + MA + Y = 0, \quad M(T) = 0.$$

Comparing with (5.20) *et seq.* we see that

$$M(t) = \int_t^T \Phi'(s, t)Y(s)\Phi(s, t)\,ds.$$

Now $Y(s)$ is non-negative definite and hence so is $M(t)$. Thus

$$\int_0^T \text{tr}(C'NC)\,dt \geqslant \int_0^T \text{tr}(C'\Lambda C)\,dt$$

and

$$\text{tr}(P_0 N(0) \geqslant \text{tr}(P_0 \Lambda(0)).$$

Thus

$$J(u^{0l}) \geqslant J(u^0)$$

as claimed. The inequality will be strict except in degenerate cases.

5.3. Partial Observations and the Separation Principle

We now consider the general partially observable regulator problem, for which the state and observation equations are:

$$dx_t = Ax_t dt + Bu(t)dt + Cdv_t$$

$$dy_t = Hx_t dt + Gdw_t. \qquad (5.38)$$

The control $\{u_t\}$ must be chosen as a function of the observation process $\{y_s, s \leqslant t\}$ so as to minimize (5.3). The question arises as to what class of control functions should be allowed. The optimal control will turn out to be $u_t = K(t)\hat{x}_t$, which is not equal to $\tilde{K}(t)y_t$ for any $\tilde{K}$, so, using a bit of hindsight, simple feedback of the latter form is insufficiently general. Staying within the linear class, a natural idea is to say that $\{u_t\}$ is admissible if it is a process such that u_t is in the linear span of the observations up to time t, i.e. if $u_t \in \mathcal{H}_t^y$ for each t. Denote by $\mathcal{V}_1$ the set of such processes. Then, according to Lemma 4.3.2, to each $u \in \mathcal{V}_1$, corresponds a function β such that

$$\sum_{ij} \int_0^t \int_0^s (\beta^{ij}(\tau, \sigma))^2 \, d\sigma d\tau < \infty \qquad (5.39)$$

and

$$u_t = \int_0^t \beta(t, s) dy_s. \qquad (5.40)$$

Substituting this in (5.38) we obtain a linear functional stochastic equation for (x_t, y_t), so that to formalize the preceding discussion it is necessary first to study the properties of such equations. This is not out of the question but is somewhat outside our present scope, particularly as an adequate finite-dimensional approximation is available. Indeed, for any function β satisfying (5.39) we can always write $\beta = \lim_m \beta_m$ where each β_m is a function of the form

$$\beta_m(t, s) = \sum_{i=1}^{k_m} K_i(t)\Delta_i(s)$$

for some matrix-valued functions $\{K_i, \Delta_i, i = 1, 2 \ldots k_m\}$. For β_m, we can easily 'realize' the relationship (5.40) in terms of a finite-dimensional linear system. Let $\Delta' = (\Delta_1, \Delta_2 \ldots \Delta_{k_m})$

and $K = (K_1, K_2 \ldots K_{k_m})$ and consider the linear system

$$dz_t = \Delta(t)dy_t, \qquad z_0 = 0$$

$$u_t = K(t)z_t.$$

Obviously

$$u_t = K(t) \int_0^t \Delta(s)dy_s$$

$$= \int_0^t \sum_i K_i(t)\Delta_i(s)dy_s$$

$$= \int_0^t \beta_m(t, s)dy_s.$$

Generalizing slightly, this leads to the following definition for a class $\mathscr{V}$ of admissible controls.

5.3.1. Definition. Let $\mathscr{V} = \{l, K(\cdot), \Gamma(\cdot), \Delta(\cdot), \zeta\}$ where l is a positive integer, K, Γ, Δ are matrices of dimensions $k \times l$, $l \times l, l \times m$ respectively, whose elements are piecewise continuous functions, and $\zeta \in R^l$. The control function corresponding to $(l, K, \Gamma, \Delta, \zeta) \in \mathscr{V}$ is

$$u_t = K(t)z_t \tag{5.41}$$

where $\{z_t\}$ is the solution of

$$dz_t = \Gamma z dt + \Delta dy_t, \qquad z_0 = \zeta. \tag{5.42}$$

We write also $u \in \mathscr{V}$ if $\{u_t\}$ is generated in this way.

To verify that this definition makes sense, note that the combined equations (5.38), (5.41), (5.42) can be written as

$$\begin{bmatrix} dx \\ dy \\ dz \end{bmatrix} = \begin{bmatrix} A & 0 & BK \\ H & 0 & 0 \\ \Delta H & 0 & \Gamma \end{bmatrix} \begin{bmatrix} x \\ y \\ z \end{bmatrix} dt + \begin{bmatrix} C & 0 \\ 0 & G \\ 0 & \Delta G \end{bmatrix} \begin{bmatrix} dv_t \\ dw_t \end{bmatrix}.$$

$$\tag{5.43}$$

which is a linear stochastic equation in the standard form, consequently having a unique solution, for any $u \in \mathscr{V}$. This is the class of controls dealt with henceforth, and it is left to the reader to determine in what sense and to what extent controls in $\mathscr{V}_1$ can be approximated by those in $\mathscr{V}$.

For $u \in \mathscr{V}$, denote the linear span of the observation process y_t^u obtained via (5.43) by

$$\mathcal{H}_t^{y,u} = \mathcal{L}\{y_s^u - Ey_s^u, s \leqslant t\}$$

and let $\mathcal{P}_t^{y,u}$ be the projection operator onto $\mathcal{H}_t^{y,u}$.

5.3.2. Proposition. $\mathcal{H}_t^{y,u}$ *does not depend on* $u \in \mathcal{V}$. *In particular*

$$\mathcal{H}_t^{y,u} = \mathcal{H}_t^{\bar{y}}$$

where $\mathcal{H}_t^{\bar{y}} = \mathcal{L}\{\bar{y}_s - E\bar{y}_s, s \leqslant t\}$, $\bar{y}_t$ *being the solution of (5.38) with* $u \equiv 0$.

Proof. This proof is very similar to the argument of Theorem 4.3.4 which shows that the innovations and observations processes span the same subspaces. Let us eliminate the 'deterministic' component by taking $m_0 = Ex_0 = 0$ and $\zeta = 0$. This does not entail any loss of generality. Note first that with $u = 0$ the state and observations are $\bar{x}_t, \bar{y}_t$ satisfying the standard Kalman filter model

$$\begin{aligned}
d\bar{x}_t &= A\bar{x}_t dt + Cdv_t & \bar{x}_0 &= 0 \\
d\bar{y}_t &= H\bar{x}_t dt + Gdw_t & \bar{y}_0 &= 0
\end{aligned} \tag{5.44}$$

Now fix $u = (l, K, \Gamma, \Delta, 0) \in \mathcal{V}$ and let x_t, y_t be as in (5.43). Define

$$x_t^* = x_t - \bar{x}_t, \qquad y_t^* = y_t - \bar{y}_t$$

By linearity, these satisfy

$$\begin{aligned}
dx_t^* &= Ax_t^* dt + Bu(t)dt & x_0^* &= 0 \\
dy_t^* &= Hx_t^* dt & y_0^* &= 0
\end{aligned} \tag{5.45}$$

Now by definition $u(t) \in \mathcal{H}_t^{y,u}$, and since (5.45) has a unique solution for given $\{u(t)\}$, evidently $x_t^*, y_t^* \in \mathcal{H}_t^{y,u}$. But

$$\bar{y}_t = y_t - y_t^*$$

and hence $\bar{y}_t \in \mathcal{H}_t^{y,u}$. Since this holds for all t,

$$\mathcal{H}_t^{\bar{y}} \subset \mathcal{H}_t^{y,u}$$

We now have to show the reverse inclusion.

Let $\Psi(t, s)$ be the $(l \times l)$ transition matrix corresponding to Γ. Then, from (5.42),

$$z_s = \int_0^s \Psi(s, \tau)\Delta(\tau)\,\mathrm{d}y_\tau$$

and hence:

$$x_t^* = \int_0^t \Phi(t, s)BKz\,\mathrm{d}s$$

$$= \int_0^t \Phi(t, s)B(s)K(s) \int_0^s \Psi(s, \tau)\Delta(\tau)\,\mathrm{d}y_\tau\,\mathrm{d}s$$

$$= \int_0^t \left(\int_\tau^t \Phi(t, s)B(s)K(s)\Psi(s, \tau)\,\mathrm{d}s \right) \Delta(\tau)\,\mathrm{d}y_\tau$$

$$= \int_0^t T(t, \tau)\,\mathrm{d}y_\tau, \text{ say.}$$

Now let $f(t) = (f^1(t), f^2(t) \ldots f^m(t))$ be a vector of functions f^i, each of which is in $L_2[0, r]$. Then since

$$\mathrm{d}\bar{y}_t = \mathrm{d}y_t - H(t)x_t^*\,\mathrm{d}t,$$

we have

$$\int_0^r f(\tau)\,\mathrm{d}\bar{y}_\tau = \int_0^r f(\tau)\,\mathrm{d}y_\tau - \int_0^r f(t)H(t) \int_0^t T(t, \tau)\,\mathrm{d}y_\tau\,\mathrm{d}t$$

$$= \int_0^r \left(f(\tau) - \int_\tau^r f(t)H(t)T(t, \tau)\,\mathrm{d}t \right)\mathrm{d}y_\tau. \quad (5.46)$$

We now invoke Lemma 4.3.3. The conditions on h are certainly satisfied, since $H(t)T(t, \tau)$ is piecewise continuous, so there exist, for $k = 1, 2, \ldots, m$, f_k such that the integrand on the right of (5.46) is equal to $e_k = (0 \ldots 0, 1, 0 \ldots 0)$. Thus

$$\int_0^r f_k(\tau)\,\mathrm{d}\bar{y}_\tau = \int_0^r e_k\,\mathrm{d}y_\tau = y_r^k.$$

This shows that $y_r^k \in \mathcal{H}_r^{\bar{y}}$ and hence that $\mathcal{H}_t^{y, u} \subset \mathcal{H}_t^{\bar{y}}$ for all t.

Let us now calculate the innovations process corresponding to (5.38). From Theorem 4.3.4 we know that $\mathcal{H}_t^{\bar{y}} = \mathcal{H}_t^\nu$ where

$$\mathrm{d}\nu_t = \mathrm{d}\bar{y}_t - H\hat{\bar{x}}_t\,\mathrm{d}t$$

and

$$\hat{\bar{x}}_t = \mathcal{P}_t^{\bar{y}}\bar{x}_t.$$

$\{\nu_t\}$ is the innovations process corresponding to (5.44). But note that

$$\hat{x}_t = \mathscr{P}_t^{y,u} x_t = \mathscr{P}_t^{\bar{y}}(\bar{x}_t + x_t^*) = \hat{\bar{x}}_t + x_t^* \qquad (5.47)$$

and hence

$$dv_t = d\bar{y}_t + (dy_t^* - Hx_t^* dt) - H\hat{\bar{x}}_t dt$$
$$= dy_t - H\hat{x}_t dt.$$

This shows that v_t is also the innovations process corresponding to

$$dy_t = Hx_t dt + Gdw_t. \qquad (5.48)$$

Now Proposition 5.3.2 says that $\mathcal{H}_t^{y,u} = \mathcal{H}_t^{\bar{y}}$. Thus $\mathcal{H}_t^{y,u} = \mathcal{H}_t^{v}$, i.e. the conclusion of Theorem 4.3.4 holds for (5.48) even though $\{x_t\}$ and $\{w_t\}$ are not orthogonal processes. (They do however satisfy condition (C).)

We can now extend to the case of non-zero $\{u_t\}$ the Kalman filtering equations for calculating $\{x_t\}$. Since (5.44) is the standard Kalman filter model, $\{\hat{\bar{x}}_t\}$ satisfies

$$d\hat{\bar{x}}_t = A\hat{\bar{x}}_t dt + PH'(GG')^{-1} dv_t$$

where $P(t)$ is given by the Riccati equation (4.40). Combining this with (5.47) and (5.45) we see that the equation generating $\hat{x}_t$ is

$$d\hat{x}_t = A\hat{x}_t dt + Bu(t)dt + PH'(GG')^{-1}(dy_t - H\hat{x}_t dt) \; (5.49)$$

which is just the standard Kalman filter equation with the control term $Bu_t dt$ added in the same way as it was in the original system equation.

The method of solving the control problem will be to convert it into a problem of 'complete observations'. Notice that since the innovations process v_t is a BM the effect of (5.49) is to replace the state equation for x_t by an equation for $\hat{x}_t$ in which all terms are 'known' (i.e. belong to $\mathcal{H}_t^{y,u}$ for each t). But the cost (5.5) is still given in terms of x_t, and the next step is to reformulate it in terms of $\hat{x}_t$. We have

$$E(x_t'Qx_t) = E((\hat{x}_t + \tilde{x}_t)'Q(\hat{x}_t + \tilde{x}_t))$$
$$= E\hat{x}_t'Q\hat{x}_t + E\tilde{x}_t'Q\tilde{x}_t + 2E\hat{x}_t'Q\tilde{x}_t$$

where

$$\tilde{x} = x - \hat{x}$$
$$= (\bar{x} + x^*) - (\hat{\bar{x}} + x^*)$$
$$= \bar{x} - \hat{\bar{x}}$$

so that $\tilde{x}$ does not depend on u, and furthermore
$E\hat{x}_t'Q\tilde{x}_t = 0$ since $\tilde{x} \perp \mathcal{H}_t^{\tilde{y}}$ and $\hat{x} \in \mathcal{H}_t^{\tilde{y}}$. Since $P_t = \text{cov}(\tilde{x}_t)$
we have from (5.19)

$$E\tilde{x}_t'Q\tilde{x}_t = \text{tr}\,\{P(t)Q(t)\}$$

Thus the cost corresponding to $u \in \mathcal{U}$ is

$$J(u) = E\left(\int_0^T (\hat{x}_t'Q\hat{x}_t + u_t'Ru_t)\mathrm{d}t + \hat{x}_T'F\hat{x}_T\right) \quad (5.50)$$

$$+ \int_0^T \text{tr}\,(P(t)Q(t))\mathrm{d}t + \text{tr}\,(P(T)F)$$

Since the last two terms do not depend on the control, it
will be seen that minimizing the original cost function is
equivalent to minimizing the same function with x replaced
by $\hat{x}$. As remarked previously, our candidate for the optimal
control is of the form $u^0 = K_1(t)\hat{x}_t$, but we first have to
show that such controls are admissible. Take any $K_1 \in \mathcal{V}$,
and define a control $u = \{l, K, \Gamma, \Delta, \zeta\}$ by

$$l = n$$

$$K = K_1$$

$$\Gamma = A - PH'(GG')^{-1}H + BK_1$$

$$\Delta = PH'(GG')^{-1}$$

$$\zeta = m_0.$$

Then from (5.42) and (5.49) the corresponding z and $\hat{x}$
processes satisfy

$$\mathrm{d}z = (A - PH'\theta H + BK_1)z\mathrm{d}t + PH'\theta\mathrm{d}y \qquad z_0 = m_0$$

$$\mathrm{d}\hat{x} = (A - PH'\theta H)\hat{x}\mathrm{d}t + BK_1 z\mathrm{d}t + PH'\theta\mathrm{d}y \qquad \hat{x}_0 = m_0,$$

where for convenience we write $\theta = (GG')^{-1}$. Thus $z - \hat{x}$
satisfies

$$\mathrm{d}(z - \hat{x}) = (A - PH'H\theta)(z - \hat{x})\mathrm{d}t \qquad z_0 - \hat{x}_0 = 0,$$

i.e. $z_t = \hat{x}_t$ for all t, so that $u_t = K_1\hat{x}_t$. Thus controls of
this form are admissible. It is now no restriction to assume
that, for any $u \in \mathcal{U}$, $\hat{x}$ is included among the components
of z, so that, changing the notation slightly, u_t can be

written as

$$u_t = K_1(t)\hat{x}_t + K_2(t)z_t = K(t)\begin{bmatrix} \hat{x}_t \\ z_t \end{bmatrix}$$

where $K = (K_1, K_2)$ and z_t is generated as before by (5.42). Recalling that from the definition of the innovations process we have

$$dy = d\nu + H\hat{x}dt, \tag{5.51}$$

equations (5.42), (5.49), (5.51) can be combined as:

$$\begin{bmatrix} d\hat{x} \\ dz \end{bmatrix} = \begin{bmatrix} A & 0 \\ \Delta H & \Gamma \end{bmatrix}\begin{bmatrix} \hat{x} \\ z \end{bmatrix}dt + \begin{bmatrix} B \\ 0 \end{bmatrix}udt + \begin{bmatrix} PH'\theta \\ \Delta \end{bmatrix}d\nu$$

i.e.

$$d\xi_t = \tilde{A}\xi_t dt + \tilde{B}u_t dt + \tilde{C}d\nu_t \tag{5.52}$$

where

$$\xi_t = \begin{bmatrix} \hat{x} \\ z \end{bmatrix}, \quad \tilde{A} = \begin{bmatrix} A & 0 \\ \Delta H & \Gamma \end{bmatrix}, \quad \tilde{B} = \begin{bmatrix} B \\ 0 \end{bmatrix}, \quad \tilde{C} = \begin{bmatrix} PH'\theta \\ \Delta \end{bmatrix}$$

and the control u_t is of the form $u_t = K\xi_t$. In terms of ξ_t, the cost is given by

$$J(u) = E\left(\int_0^T (\xi'\tilde{Q}\xi + u'Ru)dt + \xi_T'\tilde{F}\xi_T \right)$$

$$+ \int_0^T \text{tr}(PQ)dt + \text{tr}(P_T F) \tag{5.53}$$

where

$$\tilde{Q} = \begin{bmatrix} Q & 0 \\ 0 & 0 \end{bmatrix} \quad \text{and} \quad \tilde{F} = \begin{bmatrix} F & 0 \\ 0 & 0 \end{bmatrix}$$

For fixed Γ, Δ, (5.52)–(5.53) is now a standard completely observable regulator problem. From Theorem 5.2.4, the optimal control is

$$u_t^0 = -R^{-1}\tilde{B}'S\xi_t$$

where S is a solution of the Riccati equation

$$\dot{S} + S\tilde{A} + \tilde{A}'S + \tilde{Q} - S'\tilde{B}R^{-1}\tilde{B}'S = 0 \tag{5.54}$$

$$S(T) = \tilde{F}$$

Partition S similarly to F:

$$S = \begin{bmatrix} S_1 & S_2 \\ S_2' & S_3 \end{bmatrix}$$

where S_1 and S_3 are symmetric. Then (5.54) becomes

$$\begin{bmatrix} \dot{S}_1 & \dot{S}_2 \\ \dot{S}_2' & \dot{S}_3 \end{bmatrix} + \begin{bmatrix} S_1 & S_2 \\ S_2' & S_3 \end{bmatrix} \begin{bmatrix} A & 0 \\ \Delta H & \Gamma \end{bmatrix} + \begin{bmatrix} A' & H'\Delta' \\ 0 & \Gamma' \end{bmatrix} \begin{bmatrix} S_1 & S_2 \\ S_2' & S_3 \end{bmatrix}$$

$$+ \begin{bmatrix} Q & 0 \\ 0 & 0 \end{bmatrix} - \begin{bmatrix} S_1 Y S_1 & S_1 Y S_2 \\ S_2' Y S_1 & S_2 Y S_2 \end{bmatrix} = 0$$

where $Y = BR^{-1}B'$. Written out individually the three equations are:

$$(\dot{S}_1 + A'S_1 + S_1 A + Q - S_1 Y S_1) + S_2 \Delta H$$
$$+ H'\Delta'S_2' = 0 \qquad\qquad\qquad S_1(T) = F$$
$$\dot{S}_2 + S_2\Gamma + A'S_2 + S_1 Y S_2 + H'\Delta'S_3 = 0 \qquad S_2(T) = 0$$
$$\dot{S}_3 + S_3\Gamma + \Gamma'S_3 + S_2 Y S_2 = 0 \qquad\qquad S_3(T) = 0$$

Now the bracket in the first equation is the left hand side of the Riccati equation (5.31) of the completely observable problem. Thus one possible solution of the above equations is $S_1 = \Lambda, S_2 = S_3 = 0$ where $\Lambda(t)$ is the bounded symmetric solution of (5.31). The corresponding control is

$$u^0(t) = -R^{-1}\tilde{B}'S\xi_t$$
$$= -R^{-1}B'\Lambda\hat{x}_t$$

which is the same control used in the completely observable case, only with $\hat{x}_t$ replacing x_t. From (5.53) and Theorem 5.2.4, the cost corresponding to u^0 is, bearing in mind that the initial condition for (5.49) is the deterministic point m_0,

$$J(u^0) = \int_0^T \mathrm{tr}(\tilde{C}'S\tilde{C})\,ds + (m^{0'}, \zeta')S(0)\begin{bmatrix} m_0 \\ \zeta \end{bmatrix}$$

$$+ \int_0^T \mathrm{tr}(PQ)\,dt + \mathrm{tr}(P(T)F)$$

$$= \int_0^T \mathrm{tr}(\theta HP\Lambda PH'\theta)\,ds + m_0'\Lambda(0)m_0$$

$$+ \int_0^T \mathrm{tr}(PQ)\,dt + \mathrm{tr}(P(T)F)$$

But this expression does not involve Γ, Δ in any way. So we can take $\Gamma = \Delta = 0$ without affecting the cost. Note that one would not expect the solution of the Riccati equation (5.54) to be unique, because z might, for example, reproduce again the dynamics of $\hat{x}$, in which case various combinations of z and $\hat{x}$ could be used for control, all giving cost $J(u^0)$.

Summarizing, we have the following result.

5.3.3. Theorem. *For the partially observable stochastic regulator problem (5.3)–(5.5) the following control is optimal in the class $\mathscr{V}$:*

$$u_t^0 = -R^{-1}B'\Lambda\hat{x}_t$$

where $\Lambda(t)$ is the bounded symmetric solution of the Riccati equation (5.31) and $\hat{x}_t$ is the linear least squares estimate of x_t given $\{y_s, s \leq t\}$, generated by the Kalman filter (5.49). The cost corresponding to u^0 is

$$J(u^0) = m_0'\Lambda(0)m_0 + \operatorname{tr}(P(T)F)$$
$$+ \int_0^T \operatorname{tr}((GG')^{-1}HP\Lambda PH'(GG')^{-1}) + PQ)\,dt$$

where $P(t)$ is the filtering error covariance, given by the Riccati equation

$$\dot{P} = AP + PA' + CC' - PH'(GG')^{-1}HP$$
$$P_0 = \operatorname{cov}(x_0).$$

This result is known as the *separation* or *certainty-equivalence* principle and is one of the primary results in stochastic control. In general, a stochastic controller is a processing device which takes the observation record $\{y_s, s \leq t\}$ and converts it into a control value u_t. The content of Theorem 5.3.3 is that, optimally, this operation separates into two stages: computation of the statistic $\hat{x}_t$, and computation of the control value $u_t = K(t)\hat{x}_t$ as a function of this statistic. The crucial point is that these operations are independent in the sense that the Kalman filter does not depend in any way on the matrices Q, R, F defining the control problem, whereas the control function $K(t)$ does not depend on the 'noise parameters' P_0, C, G, i.e. the controller behaves as if $\hat{x}_t$ were the actual state x_t, which explains the term 'certainty-equivalence'.

Actually certainty-equivalence is a rather more special result than separation. The property that $K(t)$ is the same in the partially observable as in the completely observable case depends on the fact that, when the cost function is reformulated as in (5.50), its form as a function of $\hat{x}_t$, u_t is identical to the form of the original function of x_t, u_t. This is specifically a property of quadratic functions. On the other hand the separation result which says that the intermediate statistic to be computed is $\hat{x}_t$, regardless of the cost function in the control problem, is mainly due to the linear structure of the system dynamics, which ensures that the conditional distribution of the state is normal, and hence completely specified by $\hat{x}_t$, $P(t)$, and enables us to write x_t in the form $x_t = \bar{x}_t + x_t^*$, the parts due to noise and control respectively. For linear systems with non-quadratic costs the optimal control is of the form $u_t = \psi(t, \hat{x}_t)$, but the optimal control for the corresponding complete observations problem is some other function, not $\psi(t, x_t)$. Thus 'separation' holds, but 'certainty-equivalence' does not. However, the Ito calculus is required to prove such results; the argument is outlined in Chapter 6.

5.4. Infinite Time Problems

In many regulator applications, no specific terminal time is involved and one merely wants to ensure that the system has good 'long-run' performance. Thus we want to know what happens in the preceding cases when $T \to \infty$. Let us assume that all coefficient matrices A, B, etc., are constant (do not depend on time); then we expect that under suitable conditions the systems will settle down to some steady-state behaviour. Not surprisingly the relevant conditions hinge on the asymptotic properties of the Riccati equation. These properties are quite delicate, as evidenced by the fact that the best results are of comparatively recent vintage even though the problems are of long-standing interest, and are summarized in Theorem 5.4.2 below. The remaining results follow easily. Proofs of the assertions in Theorem 5.4.2 are not included because the methods involved are somewhat special, and because a definitive account already exists in Wonham's book [30].

Let us consider first the deterministic regulator problem:

$$\text{minimize} \int_0^T (x'(t)Qx(t) + u'(t)Ru(t))\,dt + x'(T)Fx(T)$$

$$\text{subject to } \dot{x} = Ax(t) + Bu(t), \quad x(0) = x_0 \qquad (5.55)$$

It seems at first sight that controllability of (A, B) would be necessary for long term regulation. If (A, B) is not controllable, then according to Kalman's Canonical Structure Theorem [24] it is possible to change the coordinate system in such a way that (5.55) has the form

$$\begin{bmatrix} \dot{x}^1 \\ \dot{x}^2 \end{bmatrix} = \begin{bmatrix} A_{11} & 0 \\ A_{12} & A_{22} \end{bmatrix} \begin{bmatrix} x^1 \\ x^2 \end{bmatrix} + \begin{bmatrix} 0 \\ B_2 \end{bmatrix} u$$

and x^1 cannot be regulated by the input. But there remains the possibility that x^1 might regulate itself, which it will if A_{11} is stable (all its eigenvalues have negative real parts). Pairs (A, B) with this property are said to be *stabilizable*. The following result is the reason for this terminology.

5.4.1. Proposition. *(A, B) is stabilizable if and only if there exists an $m \times n$ matrix K such that $(A + BK)$ is stable.*

Proof. See [30], Theorem 2.3.

This means that for a stabilizable system there is at least one linear feedback control $u = Kx$ such that the closed-loop system (5.55) is stable. But then

$$|x(t)| \leqslant |x_0|\,e^{-\alpha t}$$

where $-\alpha$ is the greatest real part among the eigenvalues of $A + BK$, and consequently for some constant c,

$$\int_0^\infty (x'Qx + u'Ru)\,dt = \int_0^\infty x'(Q + K'RK)x\,dt < c\,|x_0|^2.$$

Thus we can meaningfully pose the infinite-time problem

$$\text{minimize } J_\infty(u) = \int_0^\infty (x'Qx + u'Ru)\,dt \qquad (5.56)$$

$$\text{subject to} \qquad \dot{x} = Ax + Bu, \quad x(0) = x_0$$

Now factor Q as

$$Q = D'D.$$

Then we can write

$$J_\infty(u) = \int_0^\infty (|y|^2 + u'Ru)\,dt$$

where

$$y = Dx. \qquad (5.57)$$

The pair (D, A) is *unobservable* if, in some coordinate system for x, (5.55), (5.57) take the form

$$\begin{bmatrix} \dot{x}^1 \\ \dot{x}^2 \end{bmatrix} = \begin{bmatrix} A_{11} & 0 \\ A_{12} & A_{22} \end{bmatrix} \begin{bmatrix} x_1 \\ x_2 \end{bmatrix} + Bu$$

$$y = [D \quad , \quad 0] \begin{bmatrix} x^1 \\ x^2 \end{bmatrix}$$

so that x^2 does not contribute to the 'output' y. If this is the case, there could be a control u such that $J_\infty(u) < \infty$ even though $|x(t)| \to \infty$ as $t \to \infty$. However, this cannot happen if the unobservable part x^2 is stable, and pairs (D, A) with this property are called *detectable*. This is dual to stabilizability in the sense that (D, A) is detectable if and only if (A', D') is stabilizable.

These considerations suggest the reasons for the conditions in the following result on the asymptotic properties of the Riccati equation associated with the linear regulator, namely

$$\dot{\Lambda} + A'\Lambda + \Lambda A + Q - \Lambda BR^{-1}B'\Lambda = 0$$

$$\Lambda(T) = F.$$

5.4.2. Theorem. *Suppose*

$$(A, B) \text{ is stabilizable}$$
$$(D, A) \text{ is detectable} \qquad (5.58)$$
$$R > 0.$$

*Then there exists an $n \times n$ positive semidefinite matrix $\bar{\Lambda}$
such that*

(a) $\Lambda(0) \to \bar{\Lambda}$ *as* $T \to \infty$, *for any* $F \geqslant 0$

(b) $\bar{\Lambda}$ *is the unique positive semidefinite solution of the*
    algebraic Riccati equation

$$\Lambda A + A'\Lambda + Q - \Lambda BR^{-1}B'\Lambda = 0 \qquad (5.59)$$

(c) *the matrix* $A - BR^{-1}B'\bar{\Lambda}$ *is stable.*

Proof. See [30], Theorem 12.2.

Using this result we can immediately solve the infinite
time problem (5.56).

5.4.3. Theorem. *If conditions (5.58) hold, the optimal
control in the class $\mathcal{U}$ of Section 5.1 for the problem (5.56) is*

$$u^0(t) = -R^{-1}B'\bar{\Lambda}x(t)$$

with cost

$$J_\infty(u^0) = x_0'\bar{\Lambda}x_0.$$

Proof. The argument follows exactly the lines of Section 5.1
(see [30] Theorem 12.3). The important point to notice is
that, owing to the time-invariant nature of the problem, the
value function $V(t, x)$ actually does not depend on t, and
this implies that although time-varying controls $u_t = \psi(t, x(t))$
are admissible, the optimal control is time invariant, i.e.
$u_t^0 = \psi(x(t))$.

Let us now turn to the completely observable stochastic
case where the system equation is

$$dx_t = Ax\,dt + Bu\,dt + C\,dv_t.$$

In this case the cost (5.5) does not remain bounded as
$T \to \infty$. Indeed, suppose A is stabilizable and $\tilde{A} = A + BK$ is
stable. Then taking $u_t = Kx_t$ and denoting $M = Q + K'RK$
we have from Proposition 5.2.1

$$J(u) = m_0'Nm_0 + \text{tr}(P_0 N(0)) + \int_0^T \text{tr}(C'NC)\,dt.$$

where $N(t)$ is the solution of

$$\dot{N} + \tilde{A}'N + N\tilde{A} + M = 0$$

$$N(T) = F.$$

Now the transition matrix for $\tilde{A}$ is $\Phi(t, s) = e^{\tilde{A}(t-s)}$, so that, as in Section 5.2,

$$N(t) = \int_0^{T-t} e^{\tilde{A}'s} M e^{\tilde{A}s}\,ds + e^{\tilde{A}'(T-t)} F e^{\tilde{A}(T-t)}.$$

Since $\tilde{A}$ is stable, this converges as $T \to \infty$ to

$$\bar{N} = \int_0^\infty e^{\tilde{A}'s} M e^{\tilde{A}s}\,ds$$

and $\bar{N}$ is the unique solution of the algebraic equation

$$(A + BK)'N + N(A + BK) + Q + K'RK = 0.$$

It follows that $J(u) \uparrow \infty$ as $T \to \infty$, which is not unexpected since the system is constantly being perturbed by the noise $C dv_t$. So here, instead of minimizing the total cost, we will attempt to minimize the average cost per unit time:

$$J_{av}(u) = \lim_{T \to \infty} \frac{1}{T} \int_0^T (x'Qx + u'Ru)\,dt.$$

For the control $u = Kx$ as above, denote

$$\alpha(t) = \text{tr}(C'N(T-t)C).$$

Then for given $\epsilon > 0$ there is a time t, such that $|\alpha(s) - \bar{\alpha}| < \epsilon$ for all $s > t_1$, where $\bar{\alpha} = \text{tr}(C'\bar{N}C)$. Thus for $T \geq t$,

$$\left| \frac{1}{T} \int_0^T \alpha(s)\,ds - \bar{\alpha} \right| \leq \left| \frac{1}{T} \int_0^{t_1} (\alpha(s) - \bar{\alpha})\,ds \right| + \frac{1}{T} \int_{t_1}^T |\alpha(s) - \bar{\alpha}|\,ds$$

$$\leq \frac{1}{T} \left| \int_0^{t_1} (\alpha(s) - \bar{\alpha})\,ds \right| + \frac{T - t_1}{T} \epsilon.$$

Thus $1/T \int_0^T \alpha(s)\,ds \to \alpha$ as $T \to \infty$, so that

$$J_{av}(u) = \text{tr}(C'\bar{N}C).$$

This shows that $J_{av}(u)$ exists at least for constant, stabilizing controls. Let

$$\tilde{\mathscr{K}} = \{K \in \mathscr{K} : J_{av}(u) \text{ exists for } u = Kx\}$$

where $\mathscr{K}$ is as defined in Section 5.2.

5.4.4. Theorem. *If (5.58) holds, the time-invariant control* $u_t^0 = -R^{-1}B'\bar{\Lambda}x_t$ *minimizes* $J_{av}(u)$ *among all controls of the form* $u_t = Kx_t$ *with* $K \in \tilde{\mathscr{K}}$. *The minimal cost is*

$$J_{av}(u^0) = \text{tr}(C'\bar{\Lambda}C).$$

Proof. Denoting the finite-time cost for $u = Kx$ by $J_T(K)$ we have from Theorem 5.4.4

$$J_T(-R^{-1}B\Lambda) \leqslant J_T(K) \qquad \text{for all } K \in \mathscr{K}.$$

Now an argument similar to the above, using Theorem 5.4.2(a) shows that

$$\frac{1}{T}J_T(-R^{-1}B\Lambda) \to J_{av}(u^0)$$

and by assumption

$$\frac{1}{T}J_T(K) \to J_{av}(Kx)$$

if $K \in \tilde{K}$. The result follows.

Partial Observations

This case is a little more complicated as there are now two Riccati equations to be considered, and we have to use the duality between filtering and control introduced in Section 5.1 to state conditions under which both of them converge to the solutions of the respective algebraic equations.

Suppose we apply the finite time optimal control u^0 as in Theorem 5.3.3. Then the estimate $\{\hat{x}_t\}$ and error $\tilde{x}_t = x_t - \hat{x}_t$ satisfy the equations

$$d\hat{x}_t = (A - BR^{-1}B'\Lambda)\hat{x}_t dt + PH'(GG')^{-1}dv_t$$

$$d\tilde{x}_t = (A - PH'(GG')^{-1}H)\tilde{x}_t dt + Cdv_t - PH'(GG')^{-1}Gdw_t$$

where P is the solution of the Riccati equation

$$\dot{P} = AP + PA' + CC' - PH'(GG')^{-1}HP$$

$$P(0) = P_0.$$

According to the duality relations (5.16) the conditions corresponding to (5.58) are that (A', H') be stabilizable, (C', A') detectable and $GG' > 0$, which, in view of the stabilizability/detectability duality, is equivalent to:

$$(A, C) \text{ is stabilizable}$$

$$(H, A) \text{ is detectable} \qquad (5.60)$$

$$GG' > 0.$$

Under these conditions Theorem 5.4.2 asserts that $P(t) \to \bar{P}$, where $\bar{P}$ is the unique non-negative definite solution to the corresponding algebraic Riccati equation and is such that $A' - H'(GG')^{-1}H\bar{P}$ is stable. Then of course $A - \bar{P}H'(GG')^{-1}H$ is stable.

The class of admissible controls for the average cost problem will be

$$\tilde{\mathscr{V}} = \{u \in \mathscr{V} : J_{av}(u) \text{ exists}\}$$

where $\mathscr{V}$ is as defined in Section 5.3.

5.4.5. Theorem. *Suppose conditions (5.58) and (5.60) hold. Define a time-invariant control $\bar{u}^0 \in \mathscr{V}$ by*

$$\bar{u}^0 = \{n, -R^{-1}B'\bar{\Lambda}, A - BR^{-1}B'\bar{\Lambda} - PH'\theta H, PH'\theta, m_0\}.$$

Then $\bar{u}^0 \in \tilde{\mathscr{V}}$ and minimizes $J_{av}(u)$ over all $u \in \tilde{\mathscr{V}}$. Its cost is

$$J_{av}(\bar{u}^0) = \text{tr}\{(GG')^{-1}H\bar{P}\bar{\Lambda}\bar{P}H'(GG')^{-1} + \bar{P}Q\}. \quad (5.61)$$

Proof. Under conditions (5.58) and (5.60), $\Lambda(0) \to \bar{\Lambda}$ and $P(T) \to \bar{P}$ as $T \to \infty$. An argument similar to that of Theorem 5.4.4 shows that the control

$$u^0(t) = -R^{-1}B'\bar{\Lambda}\hat{x}_t$$

is in $\tilde{\mathscr{V}}$, is optimal, and has cost (5.61). But u^0 is not a time-invariant control since the Kalman filter equation (5.49) is time varying (unless P_0 happens to be equal to $\bar{P}$). However, we can use the convergence of $P(t)$, and the fact that the cost $J_{av}(u)$ does not depend on the cost over any fixed initial segment $[0, t_1]$, to show that the control $\bar{u}^0$, which is obtained by replacing $P(t)$ by $\bar{P}$ in the Kalman filter equation, achieves the same cost as u^0 and is consequently also optimal.

With $u = \bar{u}^0$ the control value is $u_t = -R^{-1}B\bar{\Lambda}z_t$, i.e. the state equation is

$$dx_t = (Ax_t - BR^{-1}B'\bar{\Lambda}z_t)dt + Cdv_t \qquad (5.62)$$

where $\{z_t\}$ is generated by

$$dz_t = (Az_t - BR^{-1}B'\bar{\Lambda}z_t)dt + \bar{P}H'\theta(dy_t - Hz_t dt). \qquad (5.63)$$

It is important to note that since $\bar{P} \neq P(t)$, z_t is not equal to $\hat{x}_t$ and in particular $dy_t - Hz_t dt$ does not have orthogonal increments. Nevertheless we have

$$dy_t - Hz_t dt = H(x_t - z_t)dt + Gdw_t$$

and hence using (5.62) and (5.63) we see that $\xi_t = x_t - z_t$ satisfies

$$d\xi_t = (A - \bar{P}H'\theta H)\xi_t dt + Cdv_t - PH'\theta Gdw_t.$$

Thus the joint process (z_t, ξ_t) satisfies an equation of the form

$$\begin{bmatrix} dz_t \\ d\xi_t \end{bmatrix} = \bar{A}\begin{bmatrix} z_t \\ \xi_t \end{bmatrix} dt + \bar{C}\begin{bmatrix} dv_t \\ dw_t \end{bmatrix}$$

where

$$\bar{A} = \begin{bmatrix} A_1 & A_2 \\ 0 & A_3 \end{bmatrix} = \begin{bmatrix} A - BR^{-1}B'\bar{\Lambda} & \bar{P}H'\theta H \\ 0 & A - \bar{P}H'\theta H \end{bmatrix}.$$

Now both A_1 and A_3 are stable, thanks to (5.58) and (5.60), and the set of eigenvalues of $\bar{A}$ is just the union of the eigenvalues of A_1 and of A_3. Thus $\bar{A}$ is stable and hence $\mathrm{cov}(z_t, \xi_t)$ converges as $t \to \infty$ to

$$\int_0^\infty e^{\bar{A}'t}\bar{C}\bar{C}'e^{\bar{A}t}dt.$$

Thus $\mathrm{cov}(x_t) = \mathrm{cov}(z_t + \xi_t)$ is convergent. This shows that $\bar{u}^0 \in \tilde{\mathcal{V}}$.

For any $u \in \tilde{\mathscr{V}}$ such that $\text{cov}(x_t)$ is convergent we have

$$J_{av}(u) = \lim_{t \to \infty} E[x_t' Q x_t + u_t' R u_t].$$

Since $P(t) \to \bar{P}$ and the covariance (5.64) is a continuous function of the elements of $\bar{A}$, we see that

$$J_{av}(u^0) = J_{av}(\bar{u}^0).$$

This completes the proof.

The average cost criterion $J_{av}(u)$ leaves something to be desired because, although it measures accurately the long-run performance, it neglects the behaviour of the system over any initial interval $[0, t_1]$, so that a wide variety of controls (e.g. any control u on $[0, t_1]$ followed by $\bar{u}^0$ on $[t_1, \infty]$) will be optimal in the sense of minimizing $J_{av}(u)$. Another possibility is to minimize the *discounted cost*

$$J_d(u) = E \int_0^\infty e^{-\rho t}(x_t' Q x_t + u_t' Q u_t) dt$$

which is finite for any stabilizing control if $\rho > 0$. This has the opposite effect: it emphasizes the initial performance. Thus one would not expect to be able to get away with a time-invariant optimal control as in the average cost case. Discounted costs are dealt with in [17].

If the stabilizability or detectability conditions are not met, a variety of things can happen; for example, the algebraic Riccati equation may have more than one non-negative definite solution. It is clear that a large number of situations can arise, depending on the exact interrelations between the various subspaces associated with the coefficient matrices A, B, etc. The reader is referred to [36] for further details. Stabilizability is a generic property in the sense of Problem 4.5.4, so we have certainly considered the most important case.

5.5. Problems and Complements

1. Define a class of controls for the completely observable regulator problem (5.24) similar to $\mathscr{V}$ of Definition 5.3.1 and

show that the minimal cost over this class is the same as that over $\mathcal{H}$ (Definition 5.2.2). Thus there is no advantage in using past-dependent controls for this problem.

2. (Correlated noise in the Kalman filter)
   Consider the system

$$dx_t = Ax_t dt + dv_t$$

$$dy_t = Hx_t dt + dw_t$$

where

$$\begin{bmatrix} v_t \\ w_t \end{bmatrix} = \int_0^t \begin{bmatrix} S(s) \\ T(s) \end{bmatrix} d\beta_s.$$

S, T, β are as in Problem 4.5.6 and x_0 is independent of $\{\beta_t\}$. Using the results of that problem to write $v_t = \hat{v}_t + \tilde{v}_t$, show that $\hat{x}_t = P_t^y x_t$ satisfies

$$d\hat{x}_t = A\hat{x}_t dt + (PH' + ST')\theta dv, \quad \hat{x}_0 = EX_0$$

where $P(t)$ is given by

$$\dot{P} = S(I - T'\theta T)S + (A - ST'\theta H)P + P(A' - H'\theta TS') - PH'\theta HP$$

$$P(0) = \text{cov}(x).$$

(I is the identity matrix and $\theta = (TT')^{-1}$.) What is the dual control problem?

3. (Control with discounted cost)
(a) Consider the problem

$$\text{minimize } J(u) = \int_0^\infty e^{-\rho t} h(x(t), u(t)) dt$$

$$\text{subject to } u(t) \in U, \quad \dot{x} = f(x, u)$$

(h bounded, $\rho > 0$). Show by proving a verification theorem similar to Theorem 5.1.1 that the Bellman equation for this problem is

$$-\rho V + \min_{u \in U} (\nabla_x V \cdot f + h) = 0.$$

Hence show that if $f = Ax + Bu$ and $h = x'Qx + u'Ru$ then the optimal control is $u(x) = -R^{-1}B'\Lambda_\rho x$, if Λ_ρ is symmetric and satisfies

$$\Lambda A + A'\Lambda - \Lambda BR^{-1}B'\Lambda - \rho\Lambda + Q = 0.$$

(b) Now consider the discounted linear regulator:

$$dx_t = Ax_t dt + Bu(t)dt + Cdv_t \quad x_0 = x$$

$$J(u) = E \int_0^\infty e^{-\rho t}(x_t'Qx_t + u_t'Ru_t)dt$$

($\rho > 0$, all matrices constant, other terms as in Section 5.2).
Assuming A is stabilizable, show from Proposition 5.2.1 that
there is at least one control $u \in \mathcal{K}$ such that $J(u)$ is finite.
What is the optimal control?

CHAPTER SIX

An outline
of further
developments

The purpose of this chapter is to provide a few signposts to
further developments in the theory of stochastic systems.
The linear/quadratic problems considered in this book are the
best developed and most used part of the subject because of
their especially simple structure. Indeed it transpires that
almost any generalization results in a substantial increase in
complication, both in terms of mathematics and of practical
implementation.

Two main generalizations have been pursued: one can drop
the assumption of linearity or of finite-dimensionality. Some
of the developments that result, for non-linear and for
infinite-dimensional systems respectively, are sketched below.
For non-linear systems the mathematics involved is mainly
probability theory. This is because mean-square methods no
longer suffice when non-linear transformations are being
considered, and one is obliged to consider in more detail the
sample function properties of the processes involved. With
infinite-dimensional linear systems, on the other hand, mean
square stochastic calculus may suffice, but the system
dynamics are specified by means of operators on function
spaces and the techniques of functional analysis play a major
role. Of course the division is not hard and fast, but for these
reasons the two developments have proceeded in somewhat
separate streams.

6.1. Non-linear Filtering and Control

As we saw in Chapter 3, Wiener integrals provide a concrete description of the linear subspaces spanned by an o.i. process $\{W_t\}$. If we want to consider non-linear filtering and control problems then Wiener integrals no longer suffice. As remarked above, it is also necessary to pay more attention to the sample function properties of the processes. Suppose $\{W_t\}$ is a BM. Then we can define the *Ito stochastic integral*

$$I_t(\omega) = \int_0^t g_s(\omega)dW_s(\omega)$$

for a certain class of *random* integrands $\{g_t\}$ in such a way that properties analogous to (3.28) hold, namely

$$E \int_0^t g_s dW_s = 0, \quad E\left[\int_0^t g_s dW_s \right]^2 = E \int_0^t g_s^2 ds \quad (6.1)$$

The basic qualification for g is that it must be *non-anticipative* with respect to $\{W_t\}$, i.e. for any $r \leqslant s \leqslant t$, $(W_t - W_s)$ must be independent of g_r. Now consider the non-linear stochastic equation

$$dX_t = m(t, X_t)dt + \sigma(t, X_t)dW_t$$
$$X_0 = \xi$$

If the functions m and σ are Lipschitz continuous in x it can be shown that this equation has a unique solution, i.e. there is just one process $\{X_t\}$ such that

$$X_t = \xi + \int_0^t m(s, X_s)ds + \int_0^t \sigma(s, X_s)dW_s$$

where the second term on the right is an Ito integral. Note that the non-anticipative condition is certainly satisfied, since $\sigma(r, X_r)$ is a function of $\{W_\tau, \tau \leqslant r\}$, and $\{W_t\}$ has independent increments.

Now suppose we want to calculate $V(t, X_t)$, where V is some suitably smooth function. Writing the Taylor series expansion (with the notation $dV(t, X_t) = V(t + dt, X_{t+dt}) - V(t, X_t)$, etc.), gives

$$dV(t, X_t) = \frac{\partial V}{\partial t}(t, X_t)dt + \frac{\partial V}{\partial x}(t, X_t)dX_t$$

$$+ \frac{1}{2}\frac{\partial^2 V}{\partial t^2}(t, X_t)(dt)^2 + \frac{1}{2}\frac{\partial^2 V}{\partial x^2}(t, X_t)(dX_t)^2$$

$$+ \ldots$$

If $\{X_t\}$ were the solution of an ordinary differential equation then only the first two terms on the right would be of first order in dt. But here X_t contains the Ito integral term $Y_t = \int_0^t \sigma(s, X_s)dW_s$, and from (6.1),

$$E(dY_t)^2 = E\sigma^2(t, X_t)dt$$

Thus the second-order term $(dX_t)^2$ contains a component which is, at least in the quadratic mean sense, of first order in dt, and which must therefore be retained in the Taylor expansion. This suggests that the correct formula for $dV(t, X_t)$ is

$$dV(t, X_t) = \frac{\partial V}{\partial t}dt + \frac{\partial V}{\partial x}dX_t + \frac{1}{2}\frac{\partial^2 V}{\partial x^2}\sigma^2 dt$$

and this, bearing in mind that $dX = m dt + \sigma dW$, is equivalent to

$$V(s, X_s) - V(r, X_r) = \int_r^s \mathcal{L}V(t, X_t)dt + \int_r^s \frac{\partial V}{\partial x}(t, X_t)\sigma(t, X_t)dW_t$$

where (6.2)

$$\mathcal{L}V(t, x) = \frac{\partial V}{\partial t}(t, x) + \frac{\partial V}{\partial x}(t, x)m(t, x)$$

$$+ \frac{1}{2}\frac{\partial^2 V}{\partial t^2}(t, x)\sigma^2(t, x).$$

This is the celebrated *Ito differential rule*. Notice that, since the stochastic integral has zero expectation,

$$E(V(s, X_s) - V(r, X_r)) = E\int_r^s \mathcal{L}V(t, X_t)dt$$

and this in effect generalizes formula (5.23) from quadratics V to any V which is differentiable once in t and twice in x. A readable account of this circle of ideas will be found in

Chapter 4 of Wong [10]. It should be stressed that the Ito differential rule depends entirely on the sample path properties of BM, and (6.2) is *not* the correct formula if $\{W_t\}$ is any other type of o.i. process.

We can apply the Ito formula (6.2) at once to control problems with dynamics of the form

$$dX_t = m(t, X_t, u_t)dt + \sigma(t, X_t)dW_t \tag{6.3}$$

and cost

$$J(u) = E\left[\int_0^T L(t, X_t, u_t)dt + g(X_T) \right] \tag{6.4}$$

We consider, as in Section 5.2, the complete observations case, so that controls will be functions $u_t = u(t, X_t)$. Then (6.3) is a stochastic differential equation. Let the value function be

$$V(t, x) = \inf_{u \mid [t,T]} E_{t,x}\left[\int_t^T L(s, X_s, u(s, X_s))ds + g(X_T) \right]$$

where the initial condition is $X_t = x$. Reasoning along the lines of Section 5.1, using also the Ito formula (6.2), suggests that V should satisfy the equation

$$\frac{\partial V}{\partial t}(t, x) + \frac{1}{2}\sigma^2(t, x)\frac{\partial^2 V}{\partial x^2}(t, x)$$

$$+ \min_u\left[\frac{\partial V}{\partial x}(t, x)m(t, x, u) + L(t, x, u) \right] = 0$$

$$V(T, x) = g(x) \tag{6.5}$$

This is the stochastic Bellman equation. It is easy to prove a verification theorem analogous to Theorems 5.1.1 and 5.2.4 which will assert that if V is a solution of (6.5) and u^0 a function such that

$$\frac{\partial V}{\partial x}(t, x)m(t, x, u^0(t, x)) + L(t, x, u^0(t, x))$$

$$= \min_u\left[\frac{\partial V}{\partial x}(t, x)m(t, x, u) + L(t, x, u) \right] \tag{6.6}$$

then u^0 is optimal. (Applied to the stochastic linear regulator of Section 5.2, this will show that a lower cost cannot be obtained with non-linear controls.) Thus the study of the control problem (6.3), (6.4) essentially consists of finding conditions on the functions m, σ, L, under which the non-linear partial differential equation (6.5) has a suitably smooth solution. It is worth remarking that the noise actually helps in this respect: it turns out that the conditions required are much less stringent when $\sigma^2 > 0$ than in the deterministic case ($\sigma = 0$). This is the main technical reason why dynamic programming is a more viable approach in stochastic than in deterministic control. A full treatment of these problems appears in Fleming and Rishel [14]. There are very few cases aside from the linear/quadratic one where a closed-form solution of the Bellman equation is available (one of them is the portfolio selection problem considered in Example 1, Section VI.4, of [14]). Usually it must be solved numerically, and the question of how to do this efficiently is a major area of research; see [17].

A complete account of *non-linear filtering* from the innovations process standpoint will be found in Liptser and Shiryaev's book [19]. Earlier accounts are [13] and [15]. The details are complicated, but the structure of the basic result is very similar to linear filtering. The problem, as before, is to estimate X_t given $\{Y_s, s \leqslant t\}$; these processes are the solutions of equations analogous to (4.20), (4.21), namely

$$dX_t = m(t, X_t)dt + \sigma(t, X_t)dV_t \qquad (6.7)$$

$$dY_t = h(t, X_t)dt + dW_t \qquad (6.8)$$

where $\{V_t\}$, $\{W_t\}$ are independent Brownian motions. We want to choose a function $U_t = U(t, \{Y_s, s \leqslant t\})$ to minimize $E(X_t - U_t)^2$. As before the appropriate function is $U_t = E[X_t | Y_s, s \leqslant t]$; see the Appendix for an outline of the way in which such conditional expectations are defined. For brevity we denote $E^t Z = E[Z | Y_s, s \leqslant t]$ for an integrable r.v. Z.

The innovations process corresponding to (6.8) is

$$d\nu_t = dY_t - E^t[h(t, X_t)] dt$$

and it transpires that this is a Brownian motion; in particular, it is normal even if $h(t, X_t)$ is not a normal process – a very striking result. The analogous result to Theorem 4.3.4 would state that $\{v_t\}$ and $\{Y_t\}$ generate the same family of σ-fields (see Appendix), which is equivalent to saying that Y_t is a function of $\{v_s, s \leqslant t\}$ and conversely. This was a long-standing conjecture which eventually turned out to be, in general, false (see [31] for an account of this circle of ideas.) Nevertheless a line of reasoning somewhat analogous to the proof of Theorem 4.4.1 leads to the conclusion that the estimate $E^t X_t$ satisfies the stochastic equation

$$d(E^t X_t) = E^t [m(t, X_t)] \, dt + E^t [X_t (h_t - E^t h_t)] \, dv_t$$
$$E^0 X_0 = E X_0 \tag{6.9}$$

where $h_t = h(t, X_t)$. The similarity between this and the Kalman filter will be clear if we take the scalar version of (4.39) with $G(t) = 1$. Then, noting that

$$E(x_t - \hat{x}_t)^2 = E[x_t (x_t - \hat{x}_t)]$$

we can write (4.39) as

$$d(\mathscr{P}_t^y x_t) = \mathscr{P}_t^y [A x_t] \, dt + E[x_t (H x_t - \mathscr{P}_t^y [H x_t])] \, dv_t$$

However, while (4.39) together with the variance equation (4.40) provide a recursive estimator for x_t, the same cannot be said for (6.9), because to update $E^t X_t$ we need to compute terms such as $E^t [m(t, X_t)]$, which are not just functions of $E^t X_t$ or of the conditional variance. Thus in non-linear filtering one has to update, not just the conditional mean, but the whole conditional distribution. The equations for this can be obtained as follows. Applying the Ito formula (6.2) to the function $V(x) = e^{iux}$ shows that e^{iuX_t} (where X_t is given by (6.7)) satisfies

$$d(e^{iuX_t}) = e^{iuX_t} [ium(t, X_t) - \tfrac{1}{2} u^2 \sigma^2 (t, X_t)] \, dt$$
$$+ iu e^{iuX_t} \sigma(t, X_t) \, dV_t. \tag{6.10}$$

Now apply the filtering formula (6.9) to the filtering problem defined by (6.10) and (6.8). This gives, written in integrated form,

$$E^t e^{iuX_t} - E e^{iuX_0} = \int_0^t E^s [e^{iuX_s}(ium(s, X_s) - \tfrac{1}{2}u^2\sigma^2(s, X_s))] \, ds$$
$$+ \int_0^t E^s [e^{iuX_s}(h(s, X_s) - E^s h(s, X_s))] \, d\nu_s \qquad (6.11)$$

This equation describes the evolution of the conditional characteristic function $E^t e^{iuX_t}$. Denote the observations $\{Y_s, s \leqslant t\}$ by η_t and suppose that the conditional distribution of X_t has a density function $p(x; t, \eta_t)$, so that

$$E^t e^{iuX_t} = \int_{-\infty}^{\infty} e^{iux} p(x; t, \eta_t) dx.$$

Then taking for example the first term in (6.11), we have

$$E^s [e^{iuX_s} ium(s, X_s)] = iu \int_{-\infty}^{\infty} e^{iux} m(s, x) p(x; s, \eta_s) dx$$

and this is the Fourier transform of

$$\frac{\partial}{\partial x} [m(s, x) p(x; s, \eta_s)]$$

Similar arguments apply to the other terms in (6.11), and taking the inverse transform of all the terms we finally obtain

$$p(x; t, \eta_t) - p_0(x) = \int_0^t \left\{ \frac{\partial}{\partial x} m(s, x) p(x; s, \eta_s) \right.$$
$$+ \frac{1}{2} \frac{\partial^2}{\partial x^2} \left. \sigma^2(s, x) p(x; s, \eta_s) \right\} ds$$
$$+ \int_0^t \left\{ h(s, x) - \int_{-\infty}^{\infty} h(s, \xi) p(\xi; s, \eta_s) d\xi \right\} p(x; s, \eta_s) d\nu_s$$

where $p_0(x)$ is the density function for x_0 and

$$d\nu_s = dY_s - \left\{ \int_{-\infty}^{\infty} h(s, \xi) p(\xi; s, \eta_s) d\xi \right\} ds.$$

These are the equations describing the evolution of the conditional density. They are essentially infinite-dimensional since in order to update $p(x; t, \eta_t)$ we need to know the whole function $p(\cdot; t, \eta_t)$. Formidable computational

problems are encountered in practical implementation of
non-linear filtering schemes based on these equations.

As regards combined filtering and stochastic control
problems, only the separation theorem has so far resulted in
solutions which are even in principle practically implement-
able. As remarked in Section 5.3, the separation result is
essentially limited to systems with linear dynamics of the
form (5.38) but can be extended to general cost functions
as in (6.4) above. The main reference here is [39]. We need
to enlarge the class of controls $\mathscr{V}$ to include non-linear
functions. The observation process $\{y_t\}$ is continuous and
admissible controls will be functions of the form
$u_t = \psi(t, y)$, where $\psi: [0, T] \times C \to R^k$, such that ψ is *non-
anticipative*, i.e. $\psi(t, y) = \psi(t, y')$ if $y(s) = y'(s)$ for $s \leqslant t$.
Then, with some restrictions on ψ, the stochastic equations
(5.38) have a unique solution. If ψ is non-linear, $\{x_t, y_t\}$
will no longer be a normal process, but the crucial point is
that the *conditional* distribution of x_t given $\{y_s, s \leqslant t\}$ is
still normal with mean given by the Kalman filter (5.49)
and variance by the Riccati equation (4.40). Let $\phi(\xi; t, x)$
denote the normal density $N(x, P(t))$ and define

$$\hat{L}(t, x, u) = \int_{R^n} L(t, \xi, u)\phi(\xi; t, x)d\xi$$

and

$$\hat{g}(x) = \int_{R^n} g(\xi)\phi(\xi; t, x)d\xi$$

Then we can write the cost (6.4) as:

$$J(u) = E\left\{ \int_0^T E[L(t, x_t, u_t) | y_s, s \leqslant t] \, dt + E[g(x_T) | y_s, s \leqslant T]\right\}$$

$$= E\left\{ \int_0^T \hat{L}(t, \hat{x}_t, u_t)dt + \hat{g}(\hat{x}_T)\right\} \tag{6.12}$$

This transforms the original partially observable problem
(5.38), (6.4) into an equivalent completely observable
problem (5.49), (6.12) which can be studied via the Bellman
equation (6.5). If a function V can be found satisfying this,
with corresponding control function $u^0(t, x)$ as in (6.6),
then the control $u_t = u^0(t, \hat{x}_t)$ will be optimal. Of course,

we still have to solve the Bellman equation to compute what u^0 actually is. Since in general $\hat{L} \neq L$ and $\hat{g} \neq g$, u^0 will not be related in any simple way to the optimal control for the equivalent completely observable problem.

In practical terms, very little headway has been made with the general non-linear partially observable problem. Intuitively it seems clear that the optimal control in such a case should be a function of the conditional distribution of the state given the observations (obtained by non-linear filtering), but even to formulate this problem in a rigorous way is hard, and the computations required to solve it seem certain to be prohibitively complex.

6.2. Distributed-parameter systems

Let us start with a specific example. Consider a rod of length l and uniform thermal conductivity κ, maintained at constant temperature 0 at each end and heated electrically by an embedded filament of resistance $c(z)$ per unit length at distance z from one end. The temperature $X(t, z)$ at time t and location z then satisfies the heat equation

$$\frac{\partial X}{\partial t} = \kappa \frac{\partial^2 x}{\partial z^2} + c(z)u(t) \qquad (6.13)$$

with boundary conditions

$$X(0, z) = X_0(z) \qquad\qquad 0 \leqslant z \leqslant 1$$
$$X(t, 0) = X(t, 1) = 0 \qquad t \geqslant 0 \qquad\qquad (6.14)$$

where X_0 is the initial temperature distribution and $u(t)$ is the square of the current in the filament at time t.[†]

The solution of (6.13), (6.14) for $u \equiv 0$ is obtained by the separation of variables technique and is

$$X(t, z) = \sum_{i=1}^{\infty} X_i(t)\phi_i(z) \qquad (6.15)$$

[†] Of course this means that as things stand $u(t)$ must be positive. However, if we regard $u(t)$ as the deviation from some constant power input then negative values are allowed.

where

and
$$\phi_i(z) = \sqrt{2}\sin i\pi z$$
$$X_i(t) = A_i e^{-\kappa i^2\pi^2 t}.$$

The coefficients A_i are chosen to satisfy the boundary condition

$$X_0(z) = X(0, z) = \sum_i A_i\phi_i(z).$$

The set $\{\phi_i\}$ is an o.n. basis for $L_2[0, 1]$ (see Proposition 2.2.6) so that for $X_0 \in L_2[0, 1]$, the A_i are given by

$$A_i = (X_0, \phi_i) = \int_0^1 X_0(z)\phi_i(z)\,dz.$$

Henceforth we denote $\lambda_i = \kappa i^2\pi^2$ and $\mathcal{L} = L_2[0, 1]$. $(\cdot, \cdot)$ and $\|\cdot\|$ always refer to the inner product and norm in $\mathcal{L}$.

Now X_i satisfies

$$\dot{X}_i = -\lambda_i X_i, \qquad X_i(0) = A_i \tag{6.16}$$

The solution of (6.13) for non-zero u can be obtained by considering the appropriate inhomogeneous version of (6.16). Indeed, suppose $c \in \mathcal{L}$ so that

$$c(z) = \sum_1^\infty c_i\phi_i(z) \tag{6.17}$$

and define $X(t, z)$ by (6.15) where X_i is now given by

$$\dot{X}_i = -\lambda_i X_i + c_i u(t), \qquad X_i(0) = A_i. \tag{6.18}$$

Differentiating formally we see that X satisfies (6.13), (6.14).

We can define the solution of these equations for certain random inputs $\{u_t\}$ in the same way as in Section 4.2. Suppose $\{V_t\}$ is a standard BM and let $\{x_t^i\}$ be the solution of the stochastic equation

Now let
$$dx_t^i = -\lambda_i x_t^i dt + c_i dV_t, \qquad x_0^i = A_i. \tag{6.19}$$

$$X_t(z) = \sum_{i=1}^\infty x_t^i\phi_i(z).$$

This *defines* the solution to (6.13), (6.14) when $\{u_t\}$ is 'white noise'. Note that

$$\text{var}\,(x_t^i) = \int_0^t e^{-2\lambda_i(t-s)} c_i^2 \, ds \leqslant \frac{c_i^2}{2\lambda_i}$$

so that

$$E\,\|X_t\|^2 = \sum_1^\infty E(x_t^i)^2 < \infty.$$

Thus $\{X_t\}$ is a well-defined process with values in $\mathcal{L}$.

Suppose we measure X_s at some point $z_0 \in (0, 1)$ over the time interval $[0, t]$ and wish to estimate the whole temperature profile $\{X_t(z), 0 \leqslant z \leqslant 1\}$. Observation strictly at a point causes difficulties of interpretation but we can formulate the more physically reasonable case where the observation is the average temperature in a small neighbourhood $(z_0 - \epsilon, z_0 + \epsilon)$, namely

$$\frac{1}{2\epsilon} \int_{z_0-\epsilon}^{z_0+\epsilon} X_s(z)\,dz = \frac{1}{2\epsilon} \sum_i x_s^i \int_{z_0-\epsilon}^{z_0+\epsilon} \phi_i(z)\,dz.$$

If this is corrupted by another independent white noise process we arrive at the observation equation

$$dY_t = \sum_i h_i x_t^i \, dt + \gamma \, dW \qquad (6.20)$$

where γ is a constant, $\{W_t\}$ a BM independent of $\{V_t\}$, and $h_i = (I_{(z_0-\epsilon, z_0+\epsilon)}, \phi_i)/2\epsilon$. (6.19) and (6.20) look very much like a Kalman filtering problem except that the state X_t takes values in the Hilbert space $\mathcal{L}$ instead of in R^n. Indeed if the function $c(z)$ happens to be a finite combination of sinusoids then we will have a finite-dimensional problem since then $c_i = 0$ for i greater than some value n, and (6.19), (6.20) can be written

$$dx_t = Ax_t dt + C dV_t$$
$$dY_t = Hx_t dt + \gamma dW_t$$

where

$$x_t = \begin{bmatrix} x_t^1 \\ \cdot \\ x_t^n \end{bmatrix}, \quad A = \begin{bmatrix} -\lambda_1 & & \\ & \ddots & \\ & & -\lambda_n \end{bmatrix}$$

$$C = \begin{bmatrix} c_1 \\ \vdots \\ c_n \end{bmatrix} \quad H' = \begin{bmatrix} h_1 \\ \vdots \\ h_n \end{bmatrix}.$$

The minimum mean-square error estimator $\hat{x}_t$ of x_t is now obtained from the Kalman filter (4.39)–(4.40), and

$$\hat{X}_t(z) = \sum_{i=1}^{n} \hat{x}_t^i \phi_i(z)$$

is therefore the best estimator in the sense that $E \|X_t - \hat{X}_t\|^2$ is minimum.

For the general problem one can consider the sequence of finite-dimensional filters obtained by truncating (6.17) to n terms, $n = 1, 2, \ldots$. One then obtains a sequence $P^n(t)$ of solutions to the $n \times n$ Riccati equations (4.40) and the question is whether this sequence converges in some sense to a family of *operators* $P(t):\mathcal{L} \to \mathcal{L}$ which will play the same role for the $\mathcal{L}$-valued process $\{X_t\}$ as $P^n(t)$ does for $\{x_t\}$, i.e. such that

$$E((X_t - \hat{X}_t), \phi_i)((X_t - \hat{X}_t), \phi_j) = (P(t) \phi_i, \phi_j)$$

where $\hat{X}_t$ is the estimate of X_t.

As in the finite-dimensional case, there is a 'dual' problem of control. Formally applying the relations (5.16) we see that this is:

$$\text{minimize} \int_0^T [(c, x(t))^2 + \gamma^2 u^2(t)] \, dt \qquad (6.21)$$

$$\text{subject to } \frac{\partial x}{\partial t} = \kappa \frac{\partial^2 x}{\partial z^2} + h(z)u(t)$$

where $u(\cdot)$ is a scalar control function and $h(z) = 1/2\epsilon$ $I_{[z_0-\epsilon, z_0+\epsilon]}(z)$. This is the same type of system as before, with the power input being proportional to $u(t)$ and uniformly distributed over the interval $[z_0 - \epsilon, z_0 + \epsilon]$, and the objective is to choose $u(\cdot)$ in order to keep the state $x(t)$ orthogonal in $\mathcal{L}$ to the fixed element c. As expected, the same operators $P(t):\mathcal{L} \to \mathcal{L}$ appear in this problem.

The questions of convergence of finite-dimensional approximations has been studied (see for example [36]) and is of course important in terms of computational methods. However, it is also possible to formulate the problems directly in a Hilbert space setting (the bible of

this subject is Lions' book [26]). Let us indicate briefly
how this is done. The initial problem is that our state space
is $\mathcal{L} = L_2[0, 1]$ but the system equation (6.13) involves
derivatives $\partial^2 X/\partial z^2$ which do not exist for all $X \in \mathcal{L}$. We
therefore introduce some subspaces of $\mathcal{L}$ on which derivatives
are defined. If f is a differentiable function on $[0, 1]$ (so
that in particular $f \in \mathcal{L}$) and ϕ is a 'smooth' function
(infinitely differentiable) with $\phi(0) = \phi(1) = 0$ then integration by parts shows that

$$\int_0^1 \frac{df}{dz}(z)\phi(z)dz = -\int_0^1 f(z)\frac{d\phi}{dz}(z)dz. \qquad (6.22)$$

We now say that a function f is weakly differentiable if there
exists a function y which plays the role of df/dz in (6.22),
i.e. such that

$$\int_0^1 y(z)\phi(z)dz = -\int_0^1 f(z)\frac{d\phi}{dz}(z)dz$$

for all smooth functions ϕ. Then we denote $y = df/dz$. Now
define

$$V = \left\{ f \in \mathcal{L} : f(0) = f(1) = 0, \right.$$

$$\left. f \text{ is weakly differentiable and } \frac{df}{dz} \in \mathcal{L} \right\}.$$

It turns out that V is a Hilbert space if given the inner product

$$(f, g)_V = (f, g) + \left(\frac{df}{dz}, \frac{dg}{dz}\right).$$

This is the simplest case of a *Sobolev space*. Now consider
the dual space V' of V, i.e. the set of continuous functions
$h : V \to R$. It is clear that every $x \in \mathcal{L}$ defines one such
function h_x by the recipe

$$h_x(f) = (x, f) \qquad (6.23)$$

so that we can regard $\mathcal{L}$ as a subset of V' (there are other
members of V' which cannot be represented by an $x \in \mathcal{L}$
as in (6.23)). So we have $V \subset \mathcal{L} \subset V'$.

To formulate the heat equation (6.13), consider the

function $a : V \times V \to R$ defined by

$$a(\phi, \psi) = \kappa \left(\frac{\mathrm{d}\phi}{\mathrm{d}z}, \frac{\mathrm{d}\psi}{\mathrm{d}z} \right).$$

For fixed ϕ, the map which takes ψ to $a(\phi, \psi)$ is an element of V'; call it A_ϕ and denote by A the linear function $A : V \to V'$ such that $A(\phi) = A_\phi$. Then we can write

$$a(\phi, \psi) = A\phi(\psi) \qquad \phi, \psi \in V.$$

Note that for smooth ϕ, ψ, integration by parts gives

$$a(\phi, \psi) = \kappa \left(\frac{\mathrm{d}\phi}{\mathrm{d}z}, \frac{\mathrm{d}\psi}{\mathrm{d}z} \right) = -\kappa \left(\frac{\mathrm{d}^2\phi}{\mathrm{d}z^2}, \psi \right)$$

so that A is an extension of the differential operator $-\kappa \mathrm{d}^2/\mathrm{d}z^2$. Consider the equation

$$\frac{\mathrm{d}X}{\mathrm{d}t} + AX = f(t), \qquad X(0) = X_0 \in \mathcal{L} \qquad (6.24)$$

where each term is, for fixed t, an element of V', to be solved on a time interval $[0, T]$. Here the derivative $\mathrm{d}X/\mathrm{d}t$ is interpreted in the weak sense as indicated earlier and f is a given square integrable function from $[0, T]$ to V'. Then a fundamental result of Lions [26, Ch. 3, Section 1.5] asserts that (6.23) has a unique solution in an appropriate sense and that if $X(t, z)$ is the solution then X and $\mathrm{d}X/\mathrm{d}t$ are in $L_2([0, 1] \times [0, T])$. Thus the effect of the above developments is to put the heat equation (6.13) in a form (6.24) which resembles the ordinary linear differential equation (5.7), although the interpretation is of course very different.

According to the above result the homogeneous equation

$$\frac{\mathrm{d}X}{\mathrm{d}t} + AX = 0 \qquad X(0) = X_0 \in \mathcal{L}$$

has a unique solution $X(t) \in \mathcal{L}$ which we can write $X(t) = G(t)X_0$. Then for each t, $G(t)$ is a continuous linear operator from $\mathcal{L} \to \mathcal{L}$, $G(0)$ is the identity and $G(t + s) = G(t)G(s)$, i.e. $\{G(t)\}$ is a *semigroup* of operators (it is the counterpart of e^{At} in the finite-dimensional case); the operator A is called the *infinitesimal generator* of $\{G(t)\}$. In this case it is clear from (6.15) that G(t) is simply given by

$$G(t)X = \sum_{i=1}^{\infty} e^{-\lambda_i t}(X, \phi_i)\, \phi_i,$$

a fact which might lead one to suppose that the introduction of all the machinery above was rather superfluous; but this is just the price we pay for illustrating a very general method by means of a very specific example.

In the finite-dimensional case the operator (matrix) A can be recovered by calculating

$$AX = -\frac{d}{dt} G(t)X \Big|_{t=0}. \tag{6.25}$$

Here this is not always possible since A is an operator from V to V' and not from $\mathcal{L}$ to $\mathcal{L}$. However, (6.25) is the correct formula on $\mathcal{D}(A) = \{X \in V : AX \in \mathcal{L}\}$ which is a dense subset of $\mathcal{L}$. Regarded as an operator from $\mathcal{D}(A) \subset \mathcal{L} \to \mathcal{L}$, A is *unbounded* (there is no constant K such that $\|AX\| \leqslant K \|X\|$ for all $X \in \mathcal{D}(A)$) and this is a major difference from linear operators on finite-dimensional spaces, which are always represented by a matrix and hence bounded.

Returning to the control problem (6.2.1) we can now state the appropriate form of Riccati equation satisfied by the operator $P(t): \mathcal{L} \to \mathcal{L}$. This is the 'weak' form in that instead of considering an equation satisfied by $P(t)$ directly, we consider the family of scalar equations:

$$\frac{d}{dt}(P(t)\phi, \psi) + (A\phi, P(t)\psi) + (P(t)\phi, A\psi)$$

$$+ \left(\left[Q - \frac{1}{\gamma^2}P(t)HH^*P(t)\right]\phi, \psi\right) = 0$$

$$P(T) = 0$$

for arbitrary $\phi, \psi \in \mathcal{D}(A)$. Here $H: R \to \mathcal{L}$ is defined by $Ha(z) = ah(z)$ and $Q: \mathcal{L} \to \mathcal{L}$ by $Q\phi(z) = (\phi, c)c(z)$. It turns out that there is just one solution $\{P(t)\}$ of this equation such that, for every $\phi, \psi \in \mathcal{L}$, $P(t)\phi$ is a continuous function of t and $(P(t)\phi, \psi)$ is absolutely continuous. The optimal control for (6.21) is related to $P(t)$ in the same way as in the finite-dimensional case. In the dual filtering problem a distributed-parameter Kalman filter is obtained using the same family of operators $\{P(t)\}$. Also, stability results for the operator Riccati equation analogous to Theorem 5.4.2 can be obtained under conditions very much analogous to the finite-dimensional ones – see [37].

In the above we have only considered the heat equation (a parabolic partial differential equation). It is possible to consider many other types of systems; as indicated earlier a major component of these problems lies in identifying the appropriate functions spaces in which to look for solutions. Another facet of such problems which has no finite-dimensional counterpart is the question of boundary behaviour. In many systems controls and/or observations are applied at the boundary (for example, at the ends of the rod considered above). In this case there is no additive forcing term in the equations but we have non-homogeneous boundary conditions. There are some delicate questions involved in formulating such problems rigorously.

In filtering, the basic reference is Bensoussan [21]. An additional problem here is how to represent 'noise' which is distributed in space. It is possible to define certain types of Hilbert-space valued noise directly. Suppose $\mathscr{X}$ is a separable Hilbert space with an o.n. basis $\{e_i\}$, $\{a_i\}$ is a sequence of numbers such that $\Sigma a_i^2 < \infty$ and $\{w_t^i\}$ are independent BM's. Then

$$w_t = \sum_{i=1}^{\infty} a_i w_t^i e_i$$

defines a $\mathscr{X}$-valued process, and there is an operator $\mathscr{W}$ not dependent on the basis $\{e_i\}$ such that for $\phi, \psi \in \mathscr{X}$,

$$E[(w_t, \phi)(w_s, \psi)] = (\mathscr{W}\phi, \psi) \cdot t \wedge s.$$

This is the $\mathscr{X}$-valued Brownian motion. Now if B is a function from $\mathscr{X}$ to $\mathcal{L}$ one can consider equations similar, for example, to (6.24) with $f = B dw_t$. This is a much more general type of noise process than is provided by the scalar BM considered earlier. The distributed-parameter Kalman filter, on the other hand, is essentially limited to finite-dimensional observations, but this is not a serious drawback from the practical standpoint

Independence and Conditional Expectation

These concepts, central to non-linear stochastic analysis, only make an appearance in this book in connection with showing that linear and non-linear estimation coincide for normal processes (Section 2.3). The general formulation of conditional expectation is usually given for integrable r.v.'s in terms of the so-called Radon–Nikodym theorem and considered a somewhat 'advanced' topic (see Section 14.3 of [4]). However, for square-integrable r.v.'s conditional expectation is just an operation of orthogonal projection and as such fits very naturally into the framework of Section 2.3.

Throughout the following, $(\Omega, \mathscr{B}, P)$ is a fixed probability space.

If $\mathscr{G}$ is a collection of sets in $\mathscr{B}$ then $\sigma(\mathscr{G})$ denotes the smallest σ-field of sets which contains $\mathscr{G}$; since $\mathscr{B}$ is a σ-field, $\sigma(\mathscr{G}) \subset \mathscr{B}$, i.e. $\sigma(\mathscr{G})$ is a *sub-σ-field* of $\mathscr{B}$. Two such collections $\mathscr{G}_1$ and $\mathscr{G}_2$ are said to be *independent* if $P(G_1 \cap G_2) = PG_1 PG_2$ for all $G_1 \in \mathscr{G}_1, G_2 \in \mathscr{G}_2$. $\mathscr{G}_1$ is *closed under finite intersections* if $G, G' \in \mathscr{G}_1$ implies $G \cap G' \in \mathscr{G}_1$.

A.1. Proposition. *Suppose $\mathscr{G}_1$ and $\mathscr{G}_2$ are closed under finite intersections and independent. Then $\sigma(\mathscr{G}_1)$ and $\sigma(\mathscr{G}_2)$ are independent.*

Proof. This is an application of the Monotone Class Theorem [4, Theorem 1.5]. Let

$$\mathscr{Q} = \{B \in \mathscr{B} : B \text{ and } G \text{ are independent, for all } G \in \mathscr{G}_2\}.$$

Then $\mathscr{Q} \supset \mathscr{G}_1$ and $\mathscr{Q}$ is a monotone class since if B_i is an increasing sequence of sets in $\mathscr{Q}$ and $B = \cup_i B_i$ then for any $G \in \mathscr{G}_2$

$$
\begin{aligned}
P(B \cap G) &= P((\underset{i}{\cup} B_i) \cap G) \\
&= P(\underset{i}{\cup}(B_i \cap G)) \\
&= \lim_i P(B_i \cap G) \\
&= \lim_i PB_i \cdot PG = PB \cdot PG.
\end{aligned}
$$

A similar argument applies to a decreasing sequence of sets. Since $\mathscr{Q}$ is a monotone class containing $\mathscr{G}_1$ it also contains $\sigma(\mathscr{G}_1)$, so that $\sigma(\mathscr{G}_1)$ and $\mathscr{G}_2$ are independent. The proof is completed by applying the same argument with $\sigma(\mathscr{G}_1)$ replacing $\mathscr{G}_2$ in the definition of $\mathscr{Q}$.

The connection between independent σ-fields and r.v.'s is as follows. Recall that a r.v. X_1 is a measurable function $X_1 : (\Omega, \mathscr{B}) \to (R, \mathscr{S})$ where $\mathscr{S}$ is the Borel σ-field of R (the σ-field generated by the class of intervals $\mathscr{S}_0 = \{[a, b) : a, b \in R, a \leqslant b\}$). This means that $\mathscr{F}_1 \subset \mathscr{B}$ where

$$\mathscr{F}_1 = X_1^{-1}(\mathscr{S}) = \{X_1^{-1}(S) : S \in \mathscr{S}\}$$

(here $X_1^{-1}(S) = \{\omega \in \Omega : X(\omega) \in S\}$). $\mathscr{F}_1$ is a σ-field; it is said to be *generated* by X_1. If X_2 is another r.v. (with generated σ-field $\mathscr{F}_2$) and the joint and marginal distribution functions of X_1, X_2 are F, F_1, F_2, then X_1 and X_2 are independent if

$$F(a_1, a_2) = F_1(a_1)F_2(a_2)$$

for all $a_1, a_2 \in R$. This implies that

$$
\begin{aligned}
&P(a_1 \leqslant X_1 < b_1, a_2 \leqslant X_2 < b_2) \\
&= P(a_1 \leqslant X_1 < b_1)P(a_2 \leqslant X_2 < b_2),
\end{aligned}
$$

i.e. the events $X_1^{-1}([a_1 b_1))$ and $X_2^{-1}([a_2, b_2))$ are independent. Thus if we let $\mathcal{G}_1 = X_1^{-1}(\mathcal{S}_0)$ and $\mathcal{G}_2 = X_2^{-1}(\mathcal{S}_0)$ then $\mathcal{G}_0$ and $\mathcal{G}_1$ are independent classes, so that by Proposition A.1, $\sigma(\mathcal{G}_1)$ and $\sigma(\mathcal{G}_2)$ are independent. But $\sigma(\mathcal{G}_1) = \mathcal{F}_1$ and $\sigma(\mathcal{G}_2) = \mathcal{F}_2$. Thus *independent r.v.'s are just r.v.'s which generate independent σ-fields.*

If $g: R \to R$ is a measurable function then the function Z defined by $Z(\omega) = g \circ X_1(\omega) = g(X_1(\omega))$ is a r.v. which has the property that

$$Z^{-1}(\mathcal{S}) \subset \mathcal{F}_1 \tag{A.1}$$

which holds because $Z^{-1}(\mathcal{S}) = X^{-1}(g^{-1}(\mathcal{S}))$. It is important that the converse is also true, namely that if Z is any r.v. satisfying (A.1) then there must be a function g such that $Z = g \circ X_1$. We can formulate this result in greater generality as follows. Suppose $(\Delta, \mathcal{D})$ is another measurable space and $Y: (\Omega, \mathcal{D}) \to (\Delta, \mathcal{D})$ a measurable function. Let $\mathcal{F} = Y^{-1}(\mathcal{D})$.

A.2. Proposition. *Let X be a r.v. Then X is $\mathcal{F}$-measurable (i.e. $X^{-1}(\mathcal{S}) \subset \mathcal{F}$) if and only if there is a measurable function $g : \Delta \to R$ such that $X(\omega) = g(Y(\omega))$.*

Proof. The 'if' part is immediate. For the converse, suppose first that X is a simple $\mathcal{F}$-measurable function; then there exist constants x_i and sets $F_i \in \mathcal{F}$ such that

$$X(\omega) = \sum_i x_i I_{F_i}(\omega).$$

Since $F_i \in \mathcal{F}$ there exist sets $D_i \in \mathcal{D}$ such that $F_i = Y^{-1}(D_i)$. We can assume that D_i are disjoint because if not we can form D_i', which are disjoint and have the same property, as follows:

$$D_1' = D_1$$

$$D_i' = D_i - \bigcup_{k=1}^{i-1} D_i' \qquad i = 2, 3 \ldots$$

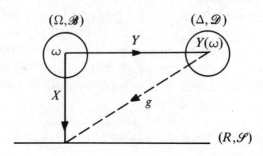

Fig. A.1

Now define $g : \Delta \to R$ as follows:

$$g(d) = \begin{cases} x_i, & d \in D_i \\ 0, & d \in (\underset{i}{\cup} D_i)^c. \end{cases}$$

Then obviously $X(\omega) = g(Y(\omega))$.

For a general $\mathscr{F}$-measurable r.v. X, let X_n be a sequence of simple functions converging to X and let g_n be the corresponding sequence of functions $g_n : \Delta \to R$ defined as above. Now let

$$g(d) = \liminf_n g_n(d).$$

Then g is $\mathscr{F}$-measurable and it is easily seen that $X(\omega) = g(Y(\omega))$.

Conditional Expectation

Let $\mathcal{H} = L_2(\Omega, \mathscr{B}, P)$ and $\mathcal{H}_0$ be defined as in Section 2.3, and let $\mathscr{F}$ be a sub-σ-field of $\mathscr{B}$. Now define

$$\mathcal{M}(\mathscr{F}) = \{Y \in \mathcal{H} : Y \text{ is } \mathscr{F}\text{-measurable}\}.$$

Then $\mathcal{M}(\mathscr{F})$ is a subspace of $\mathcal{H}$ (in fact it is just the Hilbert space $L_2(\Omega, \mathscr{F}, P)$).

A.3. Definition. *The conditional expectation of $X \in \mathcal{H}$ given $\mathscr{F}$ is the projection of X onto $\mathcal{M}(\mathscr{F})$, denoted $E^{\mathscr{F}}X$.*

According to the definition of $\mathcal{M}(\mathcal{F})$, $X \perp \mathcal{M}(\mathcal{F})$ if $E(XY) = 0$ for all $Y \in \mathcal{M}(\mathcal{F})$; however, it suffices to check that $E(XI_F) = 0$ for all $F \in \mathcal{F}$ since arbitrary Y can be approximated by linear combinations of such indicator functions. We can now establish the basic properties of conditional expectations.

A.4. Proposition. *Suppose X, $Y \in \mathcal{H}$. Then:*
(a) $E^{\mathcal{F}}(\alpha X + \beta Y) = \alpha E^{\mathcal{F}}X + \beta E^{\mathcal{F}}Y$
(b) $EX = E(E^{\mathcal{F}}X)$
(c) $E^{\mathcal{F}}X = X$ if X is $\mathcal{F}$-measurable
(d) $E^{\mathcal{F}}X = EX$ if X is independent of $\mathcal{F}$
(e) $E^{\mathcal{F}}(YX) = YE^{\mathcal{F}}X$ if $Y \in \mathcal{M}(\mathcal{F})$.

Proof. (a) and (c) are immediate from the definition, while (b) follows from the fact that $\mathcal{H}_0^1 \subset \mathcal{M}(\mathcal{F})$. For (d) we have to check that EX is the projection of X onto $\mathcal{M}(\mathcal{F})$, i.e. that $E[(X - EX)I_F] = 0$, if X and $\mathcal{F}$ are independent. But this is true because then $EXI_F = EXEI_F$. Similarly for (e) we have

$$E[(YX - YE^{\mathcal{F}}X)I_F] = E[(X - E^{\mathcal{F}}X)(YI_F)]$$
$$= 0$$

since $YI_F \in \mathcal{M}(\mathcal{F})$, so that $YE^{\mathcal{F}}X$ is the projection of YX onto $\mathcal{M}(\mathcal{F})$, as claimed.

A.5. Definition. For X, $Y \in \mathcal{H}$ the *conditional expectation of X given Y* is
$$E(X \mid Y) = E^{\mathcal{F}}X,$$
where $\mathcal{F} = Y^{-1}(\mathcal{S})$.

This says that $E(X \mid Y)$ is just the projection of X onto the subspace of r.v.'s which are measurable with respect to the σ-field generated by Y. From Proposition A.2 (with $(\Delta, \mathcal{D}) = (R, \mathcal{S})$) we conclude that $E(X \mid Y)$ is a function of the r.v. Y. If X and Y have joint density function $f(x, y)$

then it is easily checked that

$$E(X \mid Y) = \frac{\int_{-\infty}^{\infty} xf(x, Y)\,dx}{\int_{-\infty}^{\infty} f(x, Y)\,dx}$$

is the appropriate function, so that our definition coincides with the 'elementary' definition of conditional expectation. The extra generality lies in the fact that exactly the same definition applies for conditioning on much more complicated random functions Y. For example, suppose $\{Y_t\}$ is a measurable stochastic process, and let $\mathscr{F}$ be the σ-field generated by $\{Y_t\}$, defined in the following way. For any times $t_1, \ldots, t_n$ denote by $Y_{(t_1 \ldots t_n)}$ the n-vector r.v. with components $Y_{t_1} \ldots Y_{t_n}$ and by $\mathscr{S}_n$ the Borel sets of R^n. Now let

$$\mathscr{G} = \{B \in \mathscr{B} : B = Y_{(t_1 \ldots t_n)}^{-1}(S) \quad \text{for some} \quad t_1 \ldots t_n, S \in \mathscr{S}_n\}. \tag{A.2}$$

Then $\mathscr{F} = \sigma(\mathscr{G})$, and the conditional expectation of $X \in \mathcal{H}$ given $\{Y_t\}$ is just

$$E(X \mid Y_t, t \geqslant 0) = E^{\mathscr{F}}X.$$

$\{Y_t\}$ can be thought of as a random function Y whose value at $\omega \in \Omega$ is the sample function $\{Y_t(\omega), t \geqslant 0\} \in (\Delta, \mathscr{D})$, where Δ is the set of functions $f : [0, \infty) \to R$ and $\mathscr{D}$ the σ-field in Δ generated by the cylinder sets (see [4, Section 6.1]). Then $\mathscr{F} = Y^{-1}(\mathscr{D})$ so that according to Proposition A.2, $E(X \mid Y_t, t \geqslant 0)$ is a function of the process $\{Y_t\}$.

Now suppose X and $\{Y_t\}$ are jointly normal, i.e. $(X, Y_{(t_1 \ldots t_n)})$ is a normal $(n + 1)$-vector for each $t_1 \ldots t_n$. Let $\hat{X}$ be the projection of X onto $\mathcal{H}^Y = \mathcal{L}\{Y_t, t \geqslant 0\}$, and let $\tilde{X} = X - \hat{X}$. Then $\tilde{X}$ is normal from Proposition 2.3.7 and $\tilde{X} \perp \mathcal{H}^Y$ implies in particular that $\tilde{X} \perp \mathcal{L}(Y_{(t_1 \ldots t_n)})$. Since $\tilde{X}$ and $Y_{(t_1 \ldots t_n)}$ are normal and uncorrelated, they are independent. Thus $\tilde{X}$ is independent of $\mathscr{G}$ defined by (A.2). By Proposition A.1, $\tilde{X}$ and $\mathscr{F}$ are independent. Using

Proposition A.4 we see that for $F \in \mathscr{G}$,

$$
\begin{aligned}
E[I_F(X - \hat{X})] &= E[E^{\mathscr{F}}(I_F \tilde{X})] \\
&= E[I_F E^{\mathscr{F}} \tilde{X}] \\
&= E[I_F E \tilde{X}] = 0.
\end{aligned}
$$

This shows that $E^{\mathscr{F}} X = \tilde{X}$. Since $E^{\mathscr{F}} X$ is the projection onto $\mathscr{M}(\mathscr{F})$ we have

$$
E(X - E^{\mathscr{F}} X)^2 = \min_{Z \in \mathscr{M}(\mathscr{F})} E(X - Z)^2
$$

so that $E^{\mathscr{F}} X$ is the best (possible non-linear) estimator of X given $\{Y_t\}$. Thus the best linear and non-linear estimators coincide in the normal case.

References

Measure, Probability and Stochastic Processes

1 Cox, D. R. (1962), *Renewal Theory*, Chapman and Hall, London.
2 Cramér, H. and Leadbetter, M. R. (1967), *Stationary and Related Stochastic Processes*, Wiley, New York.
3 Doob, J. L. (1953), *Stochastic Processes*, Wiley, New York.
4 Kingman, J. F. C. and Taylor, S. J. (1966), *Introduction to Measure and Probability*, Cambridge University Press (first part also published separately as: S. J. Taylor, (1973), *Introduction to Measure and Integration*, Cambridge University Press.
5 Larson, R.E. (1975), *Introduction to Probability Theory and Statistical Inference*, 2nd edn, Wiley, New York.
6 P. Lévy, (1948), *Processus Stochastiques et Mouvement Brownien*, Gauthier-Villars, Paris.
7 Loève, M. (1963), *Probability Theory*, 3rd edn, D. van Nostrand, Princeton, N.J.
8 McKean, H. P. (1969), *Stochastic Integrals*, Academic Press, New York.
9 Parzen, E. (1962), *Stochastic Processes*, Holden-Day, San Francisco.
10 Wong, E. (1971), *Stochastic Processes in Information and Dynamical Systems*, McGraw-Hill, New York.

Filtering Theory and Stochastic Control

11 Balakrishnan, A. V. (1973), *Stochastic Differential Systems* I, *Lecture Notes in Economics and Mathematical Systems*, Vol. 84, Springer-Verlag, Berlin.

12 Bryson, A. E. and Ho, Y. C. (1975), *Applied Optimal Control Revised: Optimization, Estimation and Control,* Hemisphere Publishing Corp., Washington D.C.

13 Bucy, R. S. and Joseph, P. D. (1968), *Filtering for Stochastic Processes with Applications to Guidance,* Wiley Interscience, New York.

14 Fleming, W. H. and Rishel, R. W. (1975), *Deterministic and Stochastic Optimal Control,* Springer-Verlag, New York.

15 Jazwinski, A. H. (1970), *Stochastic Processes and Filtering Theory,* Academic Press, New York.

17 Kushner, H. J. (1977), *Probability Methods for Approximations in Stochastic Control and for Elliptic Equations,* Academic Press, New York.

18 Kwakernaak, H. and Sivan, R. (1972), *Linear Optimal Control Systems,* Wiley Interscience, New York.

19 Liptser, R. S. and Shiryaev, A. N. (1974), *Statistics of Stochastic Processes,* Nauka, Moscow (in Russian; English translation to be published by Springer-Verlag).

20 Wonham, W. M. (1967), *Lecture Notes in Stochastic Control,* Center for Dynamical Systems, Brown University, Providence, R.I.

21 Bensoussan, A. (1971), *Filtrage Optimal des Systèmes Linéaires,* Dunod, Paris.

Other Books

22 Brockett, R. W. (1970), *Finite-Dimensional Linear Systems,* Wiley, New York.

23 Cochran, J. A. (1972), *The Analysis of Linear Integral Equations,* McGraw-Hill, New York.

24 Desoer, C. A. (1970), *Notes for a Second Course on Linear Systems,* Van Nostrand Reinhold, New York.

25 Ferguson, T. S. (1967), *Mathematical Statistics: a Decision-Theoretic Approach,* Academic Press, New York.

26 Lions, J. L. (1971), *Optimal Control of Systems Governed by Partial Differential Equations,* Springer-Verlag, Berlin.

27 Raiffa, H. and Schlaifer, R. (1961), *Applied Statistical Decision Theory,* MIT Press, Cambridge, Mass.

28 Taylor, A. E. (1958), *Introduction to Functional Analysis,* Wiley, New York.

29 Van Trees, H. (1968), *Detection, Estimation and Modulation Theory,* Part I, Wiley, New York.

30 Wonham, W. M. (1974), *Linear Multivariable Control, a Geometric Approach, Lecture Notes in Economics and Mathematical Systems,* Vol. 101, Springer-Verlag, Berlin.

Journal Articles

31 Beneš, V. E. (1976), On Kailath's Innovations Conjecture, *Bell System Technical Journal*, **55**, 981–1001.

32 Kailath, T. (1968), An innovations approach to least-squares estimation. Part I: Linear filtering in additive white noise, *IEEE Trans. Automatic Control*, **AC-13**, 646–654.

33 Kailath, T. and Frost, P. A. (1968), An innovations approach to least-squares estimation. Part II: Linear smoothing in additive white noise, *IEEE Trans. Automatic Control*, **AC-13**, 655–660.

34 Kalman, R. E. (1960), A new approach to linear filtering and prediction problems, *ASME Transactions*, Part D (*J. of Basic Engng*) **82**, 35–45.

35 Kalman, R. E. and Bucy, R. S. (1961), New results in linear filtering and prediction theory, *ASME Transactions*, **83**, Part D (*J. of Basic Engng*) 95–108.

36 Lukes, D. L. and Russell, D. L. (1969), The quadratic criterion for distributed systems, *SIAM J. Control*, **7**, 101–121.

37 Vinter, R. B. (1977), Filter stability for stochastic evolution equations, *SIAM J. Control and Opt.*, **15**, 465–485.

38 Willems, J. C. (1971), Least squares stationary optimal control and the algebraic Riccati equation, *IEEE Trans. Automatic Control*, **AC-16**, (6), 621–634.

39 Wonham, W. M. (1968), On the separation theorem of stochastic control, *SIAM J. Control*, **6**, 312–326.

Index